Composite Farming Practices and Economic Development

Composite Farming Practices and Economic Development

Prof. Sriniwas Puri

RANDOM

RANDOM PUBLICATIONS

NEW DELHI (INDIA)

Composite Farming Practices and Economic Development

ISBN 978-93-5111-590-8

Published in 2015 in India by

RANDOM PUBLICATIONS

Reprint, 2017

4376-A/4B, Gali Murari Lal, Ansari Road
New Delhi-110 002
Phone : +9111-43580356, 011-23289044, 011-43142548
e-mail: sales@randompublications.com,
info@randompublications.com, randomexports@gmail.com

Type Setting by : Friends Media, Delhi-110089
Digitally Printed at : Replika Press Pvt. Ltd.

Preface

Since agriculture constitutes a large share of national output and employs a majority of the labor force in most developing countries, the sector has been integral to any thinking about development. However, the perceived role of agriculture in growth and development has changed considerably over the past half-century. Early classical theory viewed economic development as a growth process requiring the systematic reallocation of factors of production from a primary sector characterized by low productivity, traditional technology, and decreasing returns to a modern industrial sector with higher productivity and increasing returns. Composit Farming Practices and Economic Development Agriculture and allied activities sector is the most important segment of the economy. Technological progress and introduction of Scientific agricultural inputs have made remarkable improvements in agriculture and allied activities. It is technological innovation and eforts of devoted scientists, which enables us to achieve strategic breakthrough in agriculture. The process of diversification in agriculture, the success of technological breakthrough is area spacifc as well as crop specific. Some other emerging problems in agricultural sector are-soil degradation, environmental pollution, food security, non sustainable agricultural practice, credit availability etc.

I would like to thank my team for standing beside me throughout my career and writing this book. My special thanks go to "Random Publications" who have published the book.

– Prof. Sriniwas Puri

Contents

1

Farming Methods and Principles

AN OVERVIEW

Farming methods combine scientific knowledge of ecology and modern technology with traditional farming practices based on naturally occurring biological processes. Farming methods are studied in the field of agroecology. While conventional agriculture uses synthetic pesticides and water-soluble synthetically purified fertilizers, organic farmers are restricted by regulations to using natural pesticides and fertilizers. The principal methods of farming include crop rotation, green manures and compost, biological pest control, and mechanical cultivation.

These measures use the natural environment to enhance agricultural productivity: legumes are planted to fix nitrogen into the soil, natural insect predators are encouraged, crops are rotated to confuse pests and renew soil, and natural materials such as potassium bicarbonate and mulches are used to control disease and weeds. Organic farmers are careful in their selection of plant breeds, and organic researchers produce hardier plants through plant breeding rather than genetic engineering.

CROP DIVERSITY

Crop diversity is a distinctive characteristic of farming. Conventional farming focuses on mass production of one crop in one location, a practice called monoculture. This makes apparent economic sense: the larger the growing area, the lower the per unit cost of fertilizer, pesticides and specialized machinery for a single plant species. The science of agroecology has revealed the benefits of polyculture which is often employed in farming. Planting a variety of vegetable crops supports a wider range of beneficial insects, soil microorganisms, and other factors that add up to overall farm health, but managing the balance requires expertise and close attention.

FARM SIZE

Farm size in great measure determines the general approach and specific tools and methods. Today, major food corporations are involved in all aspects of organic production on a large scale. However, farming originated as a small-scale enterprise, with operations from under 1-acre to under 100 acres.

The mixed vegetable organic market garden is often associated with fresh, locally-grown produce, farmers' markets and the like, and this type of farm is often under 10 acres. Farming at this scale is generally Labour-intensive, involving more manual Labour and less mechanization. The type of crop also determines size: organic grain farms often involve much larger area. Larger organic farms tend to use methods and equipment similar to conventional farms, centered around the tractor.

PLANT NUTRITION

Soil Fertility

The central farming activity of fertilization Emphasizes the differences. Farming relies heavily on the natural breakdown of organic matter, using techniques like green manure and composting, to replace nutrients taken from the soil. This biological process, driven by microorganisms such as mycorrhiza, allows the natural production of nutrients in the soil throughout the growing season, and has been referred to as feeding the soil to feed the plant. In chemical farming, individual nutrients, like nitrogen, are synthesized in a more or less pure form that plants can use immediately, and applied on a man-made schedule. Each nutrient is defined and addressed separately. Problems that may arise from one action are usually addressed with additional, corrective products and procedures.

Farming uses a variety of methods to improve soil fertility, including crop rotation, cover cropping, and application of compost.

PEST CONTROL

Differing approaches to pest control are equally notable. In chemical farming, a specific insecticide may be applied to quickly kill off a particular insect pest. Chemical controls can dramatically reduce pest populations for the short term, yet by unavoidably killing natural predator insects and animals, cause an ultimate increase in the pest population. Repeated use of insecticides and herbicides and other pesticides also encourages rapid natural selection of resistant insects, plants and other organisms, necessitating increased use, or requiring new, more powerful controls.

In contrast, farming tends to tolerate some pest populations while taking a longer-term approach.

Organic pest control involves the cumulative effect of many techniques, including:

- Allowing for an acceptable level of pest damage;

- Encouraging predatory beneficial insects to control pests;
- Encouraging beneficial microorganisms and insects; this by serving them nursery plants and/or an alternative habitat, usually in a form of a shelterbelt, hedgerow, or beetle bank;
- Careful crop selection, choosing disease-resistant varieties;
- Planting companion crops that discourage or divert pests;
- Using row covers to protect crops during pest migration periods;
- Using pest regulating plants and biologic pesticides and herbicides;
- Using no-till farming, and no-till farming techniques as false seedbeds ;
- Rotating crops to different locations from year to year to interrupt pest reproduction cycles; and
- Using insect traps to monitor and control insect populations.

Each of these techniques also provides other benefits—soil protection and improvement, fertilization, pollination, water conservation, season extension, etc.—and these benefits are both complementary and cumulative in overall effect on farm health. Effective organic pest control requires a thorough understanding of pest life cycles and interactions. Organic pest control is similar to integrated pest management in some respects.

LIVESTOCK

Raising livestock and poultry, for meat, dairy and eggs, is another traditional, farming activity that complements growing. Organic farms attempt to provide animals with "natural" living conditions and feed. While the USDA does not require any animal welfare requirements be met for a product to be marked as organic, this is a variance from older farming practices. Ample, free-ranging outdoor access, for grazing and exercise, is a distinctive feature, and crowding is avoided. Feed is also organically grown, and drugs, including antibiotics, are not ordinarily used. Animal health and food quality are thus pursued in a holistic "fresh air, exercise, and good food" approach.

Also, horses and cattle used to be a basic farm feature that provided Labour, for hauling and plowing, fertility, through recycling of manure, and fuel, in the form of food for farmers and other animals. While today, small growing operations often do not include livestock, domesticated animals are a desirable part of the farming equation, especially for true sustainability, the ability of a farm to function as a self-renewing unit.

FARMING SYSTEMS

There are several farming systems. Biodynamic farming is a comprehensive approach, with its own international governing body. The Do Nothing Farming method focuses on a minimum of mechanical cultivation and Labour for grain crops. French intensive and biointensive, methods are well-suited to organic principles. Other techniques are permaculture and no-till farming. Finally, newcomers as the Agro-ecologic system focus on a blend of a more large-scale approach with imbedded natural/farming techniques. A farm may choose to adopt a particular method, or a mix of techniques. While fundamentally different, large-scale agriculture and farming are not entirely mutually exclusive. Integrated Pest Management is a multifaceted strategy that can include synthetic pesticides as a last resort—both organic and conventional farms use IPM systems for pest control.

SUSTAINABLE AGRICULTURE

Sustainable agriculture is a special kind of agriculture along with farming technique which makes maximum utilization of the environment without causing any form of harm to it. The outputs from this Sustainable Agriculture are devoid of any types of inorganic chemicals such as pesticides and insecticides. This farming technique uses the organic way of farming.

All these factors produce the products in a more environment friendly manner and are thus healthier for the consumers to consume.

- Sustainable Agriculture Practice is regarded as the need of the hour by most of the countries over the world keeping in view environmental concern.
- Sustainable Agriculture Washington is an organization, which has been rendering promotional methods in the field of agriculture. Find more about Sustainable Agriculture in Washington.
- National Sustainable Agriculture is one of the leading national level Programme in United States of America. Get more on National Sustainable Agriculture.

The other characteristic feature of Sustainable Agriculture are given in the bullet form below:

- The treatment towards the animals in such kind of farming are more humane because they are given enough space to graze and are not bred like factory products.
- The workers who are associated with the Sustainable Agriculture receive more humane Behaviour from the employers and it has been observed that the wages offered to these farm workers are fairer than the ones working in the so called industrial farms.
- The one of the most important feature of Sustainable Agriculture is its extended support towards the small farming community farmers. It is essentially being emphasized in this type of farming that the organic food items that have been produced by the small farmers of the rural areas are bought and marketed to the consumers through this Sustainable Agriculture route.

Sustainable Agriculture got its first legal recognition in the United States of America through the Public Law 101-624, Title XVI, Subtitle A, Section 1683 which was introduced in the year 1990. This law defined this farming practice type to be more Eco-friendly and having a sight for the long run satisfaction of the nutrient values in accordance with the nutritional requirements of the human body. It was especially emphasized in the law that the Non- renewable resources that are used in the farming process needs to be used at a minimal level in Sustainable Agriculture. It emphasizes that the viability conditions of the small farmers in the rural areas needs to be enhanced through Sustainable Agriculture through proper disposal of wage and proper distribution of their produced products.

Information on Sustainable Agriculture would be incomplete if the mention of the Farming is not considered in the same. For more development in this field, research on sustainable agriculture is going on and with every passing day, this farming type is given more emphasis by the American as well as the French government.

CONCEPT THEMES

Sustainable agriculture integrates three main goals—environmental health, economic profitability, and social and economic equity. A variety of philosophies, policies and practices have contributed to these goals. People in many different capacities, from farmers to consumers, have shared this vision and contributed to it. Despite the diversity of people and perspectives, the following themes commonly weave through definitions of sustainable agriculture.

Sustainability rests on the principle that we must meet the needs of the present without compromising the ability of future generations to meet their own needs. Therefore, *stewardship of both natural and human resources* is of prime importance. Stewardship of human resources includes consideration of social responsibilities such as working and living conditions of laborers, the needs of rural communities, and consumer health and safety both in the present and the future. Stewardship of land and natural resources involves maintaining or enhancing this vital resource base for the long term.

A *systems perspective* is essential to understanding sustainability. The system is envisioned in its broadest sense, from the individual farm, to the local ecosystem, *and* to communities affected by this farming system both locally and globally. An emphasis on the system allows a larger and more thorough view of the consequences of farming practices on both human communities and the environment. A systems approach gives us the tools to explore the interconnections between farming and other aspects of our environment. A systems approach also implies *interdisciplinary efforts in research and education*. This requires not only the input of researchers from various disciplines, but also farmers, farmworkers, consumers, policymakers and others.

Making the transition to sustainable agriculture is a process. For farmers, the transition to sustainable agriculture normally requires a series of small, realistic steps. Family economics and personal goals influence how fast or how far participants can go in the transition. It is important to realize that each small decision can make a difference and contribute to advancing the entire system further on the "sustainable agriculture continuum." The key to moving forward is the will to take the next step.

Finally, it is important to point out that *reaching toward the goal of sustainable agriculture is the responsibility of all participants in the system,* including farmers, laborers, policymakers, researchers, retailers, and consumers. Each group has its own part to play, its own unique contribution to make to strengthen the sustainable agriculture community.

The remainder of this document considers specific strategies for realizing these broad themes or goals. The strategies are grouped according to three separate though related areas of concern: Farming and Natural Resources, Plant and Animal Production Practices, and the Economic, Social and Political Context. They represent a range of potential ideas for individuals committed to interpreting the vision of sustainable agriculture within their own circumstances.

FARMING AND NATURAL RESOURCES

Water. When the production of food and Fibre degrades the natural resource base, the ability of future generations to produce and flourish decreases. The decline of ancient civilizations in Mesopotamia, the Mediterranean region, Pre-Columbian southwest U.S. and Central America is believed to have been strongly influenced by natural resource degradation from non-sustainable farming and forestry practices. Water is the principal resource that has helped agriculture and society to prosper, and it has been a major limiting factor when mismanaged.

Water supply and use. In California, an extensive water storage and transfer system has been established which has allowed crop production to expand to very arid regions. In drought years, limited surface water supplies have prompted overdraft of groundwater and consequent intrusion of salt water, or permanent collapse of aquifers. Periodic droughts, some lasting up to 50 years, have occurred in California.

Several steps should be taken to develop drought-resistant farming systems even in "normal" years, including both policy and management actions:

- Improving water conservation and storage measures,
- Providing incentives for selection of drought-tolerant crop species,
- Using reduced-volume irrigation systems,
- Managing crops to reduce water loss, or
- Not planting at all.

Water quality: The most important issues related to water quality involve salinization and contamination of ground and surface waters by pesticides, nitrates and selenium. Salinity has become a problem wherever water of even relatively low salt content is used on shallow soils in arid regions and/or where the water table is near the root zone of crops. Tile drainage can remove the water and salts, but the disposal of the salts and other contaminants may negatively affect the environment depending upon where they are deposited.

Temporary solutions include the use of salt-tolerant crops, low-volume irrigation, and various management techniques to minimize the effects of salts on crops. In the long-term, some farmland may need to be removed from production or converted to other uses. Other uses include conversion of row crop land to production of drought-tolerant forages, the restoration of wildlife habitat or the use of agroforestry to minimize the impacts of salinity and high water tables. Pesticide and nitrate contamination of water can be reduced using many of the practices discussed later in the Plant Production Practices.

Wildlife: Another way in which agriculture affects water resources is through the destruction of riparian habitats within watersheds. The conversion of wild habitat to agricultural land reduces fish and wildlife through erosion and sedimentation, the effects of pesticides, removal of riparian plants, and the diversion of water. The plant diversity in and around both riparian and agricultural areas should be maintained in order to support a diversity of wildlife. This diversity will enhance natural ecosystems and could aid in agricultural pest management.

Energy: Modern agriculture is heavily dependent on non-renewable energy sources, especially petroleum. The continued use of these energy sources cannot be sustained indefinitely, yet to abruptly abandon our reliance on them would be economically catastrophic. However, a sudden cutoff in energy supply would be equally disruptive. In sustainable agricultural systems, there is reduced reliance on non-renewable energy sources and a substitution of renewable sources or Labour to the extent that is economically feasible.

Air: Many agricultural activities affect air quality. These include smoke from agricultural burning; dust from tillage, traffic and harvest; pesticide drift from spraying; and nitrous oxide emissions from the use of nitrogen fertilizer. Options to improve air quality include incorporating crop residue into the soil, using appropriate levels of tillage, and planting wind breaks, cover crops or strips of native perennial grasses to reduce dust.

Soil: Soil erosion continues to be a serious threat to our continued ability to produce adequate food. Numerous practices have been developed to keep soil in place, which include reducing or eliminating tillage, managing irrigation to reduce Run-off.

PLANT PRODUCTION PRACTICES

Sustainable production practices involve a variety of approaches. Specific strategies must take into account topography, soil characteristics, climate, pests, local availability of inputs and the individual grower's goals.

Despite the site-specific and individual nature of sustainable agriculture, several general principles can be applied to help growers select appropriate management practices:

- Selection of species and varieties that are well suited to the site and to conditions on the farm;
- Diversification of crops and cultural practices to enhance the biological and economic stability of the farm;
- Management of the soil to enhance and protect soil quality;
- Efficient and humane use of inputs; and
- Consideration of farmers' goals and lifestyle choices.

Selection of site, species and variety: Preventive strategies, adopted early, can reduce inputs and help establish a sustainable production system. When possible, pest-resistant crops should be selected which are tolerant of existing soil or site conditions. When site selection is an option, factors such as soil type and depth and location should be taken into account before planting.

Diversity: Diversified farms are usually more economically and ecologically resilient. While monoculture farming has advantages in terms of efficiency and ease of management, the loss of the crop in any one year could put a farm out of business and/or seriously disrupt the stability of a community dependent on that crop. By growing a variety of crops, farmers spread economic risk and are less susceptible to the radical price fluctuations associated with changes in supply and demand.

Properly managed, diversity can also buffer a farm in a biological sense. In annual cropping systems, crop rotation can be used to suppress weeds, pathogens and insect pests. Also, cover crops can have stabilizing effects on the agroecosystem by holding soil and nutrients in place, conserving soil moisture with mowed or standing dead mulches, and by increasing the water infiltration rate and soil water holding capacity. Cover crops in orchards and vineyards can buffer the system against pest infestations by increasing beneficial arthropod populations and can therefore reduce the need for chemical inputs. Using a variety of cover crops is also important in order to protect against the failure of a particular species to grow and to attract and sustain a wide range of beneficial arthropods.

Optimum diversity may be obtained by integrating both crops and livestock in the same farming operation. This was the common practice for centuries until the mid-1900s when technology, government policy and economics compelled farms to become more specialized. Mixed crop and livestock operations have several advantages. First, growing row crops only on more level land and pasture or forages on steeper slopes will reduce soil erosion. Second, pasture and forage crops in rotation enhance soil quality and reduce erosion; livestock manure, in turn, contributes to soil fertility. Third, livestock can buffer the negative impacts of low rainfall periods by consuming crop residue that in "plant only" systems would have been considered crop failures. Finally, feeding and marketing are flexible in animal production systems. This can help cushion farmers against trade and price fluctuations and, in conjunction with cropping operations, make more efficient use of farm Labour.

Soil management: A common philosophy among sustainable agriculture practitioners is that a "healthy" soil is a key component of sustainability; that is, a healthy soil will produce healthy crop plants that have optimum Vigour and are less susceptible to pests. While many crops have key pests that attack even the healthiest of plants, proper soil, water and nutrient management can help prevent some pest problems brought on by crop stress or nutrient imbalance. Furthermore, crop management systems that impair soil quality often result in greater inputs of water, nutrients, pesticides, and/or energy for tillage to maintain yields.

In sustainable systems, the soil is viewed as a fragile and living medium that must be protected and nurtured to ensure its long-term productivity and stability. Methods to protect and enhance the productivity of the soil include using cover crops, compost and/or manures, reducing tillage, avoiding traffic on wet soils, and maintaining soil cover with plants and/or mulches. Conditions in most California soils do not Favour the buildup of organic matter. Regular additions of organic matter or the use of cover crops can increase soil aggregate stability, soil tilth, and diversity of soil microbial life.

Efficient use of inputs: Many inputs and practices used by conventional farmers are also used in sustainable agriculture. Sustainable farmers, however, maximize reliance on natural, renewable, and on-farm inputs. Equally important are the environmental, social, and economic impacts of a particular strategy. Converting to sustainable practices does not mean simple input substitution. Frequently, it substitutes enhanced management and scientific knowledge for conventional inputs, especially chemical inputs that harm the environment on farms and in rural communities. The goal is to develop efficient, biological systems which do not need high levels of material inputs.

Growers frequently ask if synthetic chemicals are appropriate in a sustainable farming system. Sustainable approaches are those that are the least toxic and least energy intensive, and yet maintain productivity and profitability. Preventive strategies and other alternatives should be employed before using chemical inputs from any source.

However, there may be situations where the use of synthetic chemicals would be more "sustainable" than a strictly non-chemical approach or an approach using toxic "organic" chemicals. One grape grower switched from tillage to a few applications of a broad spectrum contact herbicide in the vine row. This approach may use less energy and may compact the soil less than numerous passes with a cultivator or mower. Consideration of farmer goals and lifestyle choices. Management decisions should reflect not only environmental and broad social considerations, but also individual goals and lifestyle choices. Adoption of some technologies or practices that promise profitability may also require such intensive management that one's lifestyle actually deteriorates. Management decisions that promote sustainability, nourish the environment, the community *and* the individual.

ANIMAL PRODUCTION PRACTICES

In the early part of this century, most farms integrated both crop and livestock operations. Indeed, the two were highly complementary both biologically and economically. The current picture has changed quite drastically since then. Crop and animal producers now are still dependent on one another to some degree, but the integration now most commonly takes place at a higher level—*between* farmers, through intermediaries, rather than *within* the farm itself.

This is the result of a trend Towards separation and specialization of crop and animal production systems. Despite this trend, there are still many farmers, particularly in the Midwest and Northeastern U.S. that integrate crop and animal systems—either on dairy farms, or with range cattle, sheep or hog operations.

Even with the growing specialization of livestock and crop producers, many of the principles outlined in the crop production section apply to both groups. The actual management practices will, of course, be quite different. Some of the specific points that livestock producers need to address are listed below.

Management Planning: Including livestock in the farming system increases the complexity of biological and economic relationships. The mobility of the stock, daily feeding, health concerns, breeding operations, seasonal feed and forage sources, and complex marketing are sources of this complexity. Therefore, a successful ranch plan should include enterprise calendars of operations, stock flows, forage flows, Labour needs, herd production records and land use plans to give the manager control and a means of monitoring progress Towards goals.

Animal Selection: The animal enterprise must be appropriate for the farm or ranch resources. Farm capabilities and constraints such as feed and forage sources, landscape, climate and skill of the manager must be considered in selecting which animals to produce. Ruminant animals can be raised on a variety of feed sources including range and pasture, cultivated forage, cover crops, shrubs, weeds, and crop residues. There is a wide range of breeds available in each of the major ruminant species, *i.e.*, cattle, sheep and goats. Hardier breeds that, in general, have lower growth and milk production potential, are better adapted to less favorable environments with sparse or highly seasonal forage growth.

Animal nutrition: Feed costs are the largest single variable cost in any livestock operation. While most of the feed may come from other enterprises on the ranch, some purchased feed is usually imported from off the farm. Feed costs can be kept to a minimum by monitoring animal condition and performance and understanding seasonal variations in feed and forage quality on the farm. Determining the optimal use of farm-generated by-products is an important challenge of diversified farming.

Reproduction: Use of quality germplasm to improve herd performance is another key to sustainability. In combination with good genetic stock, adapting the reproduction season to fit the climate and sources of feed and forage reduce health problems and feed costs.

Herd Health: Animal health greatly influences reproductive success and weight gains, two key aspects of successful livestock production. Unhealthy stock waste feed and require additional Labour. A herd health Programme is critical to sustainable livestock production.

Grazing Management: Most adverse environmental impacts associated with grazing can be prevented or mitigated with proper grazing management. First, the number of stock per unit area must be correct for the landscape and the forage sources. There will need to be compromises between the convenience of tilling large, unfenced fields and the fencing needs of livestock operations.. Second, the long term carrying capacity and the stocking rate must take into account short and long-term droughts. Especially in Mediterranean climates such as in California, properly managed grazing significantly reduces fire hazards by reducing fuel build-up in grasslands and brushlands. Finally, the manager must achieve sufficient control to reduce overuse in some areas while other areas go unused. Prolonged concentration of stock that results in permanent loss of vegetative cover on uplands or in riparian zones should be avoided. However, small scale loss of vegetative cover around water or feed troughs may be tolerated if surrounding vegetative cover is adequate.

Confined Livestock Production: Animal health and waste management are key issues in confined livestock operations. The moral and ethical debate taking place today regarding animal welfare is particularly intense for confined livestock production systems. The issues raised in this debate need to be addressed.

Confinement livestock production is increasingly a source of surface and ground water pollutants, particularly where there are large numbers of animals per unit area. Expensive waste management facilities are now a necessary cost of confined production systems. Waste is a problem of almost all operations and must be managed with respect to both the environment and the quality of life in nearby communities. Livestock production systems that disperse stock in pastures so the wastes are not concentrated and do not overwhelm natural nutrient cycling processes have become a subject of renewed interest.

ECONOMIC, SOCIAL AND POLITICAL CONTEXT

In addition to strategies for preserving natural resources and changing production practices, sustainable agriculture requires a commitment to changing public policies, economic institutions, and social values. Strategies for change must take into account the complex, reciprocal and ever-changing relationship between agricultural production and the broader society.

The "food system" extends far beyond the farm and involves the interaction of individuals and institutions with contrasting and often competing goals including farmers, researchers, input suppliers, farmworkers, unions, farm advisors, processors, retailers, consumers, and policymakers. Relationships among these actors shift over time as new technologies spawn economic, social and political changes.

A wide diversity of strategies and approaches are necessary to create a more sustainable food system. These will range from specific and concentrated efforts to alter specific policies or practices, to the longer-term tasks of reforming key institutions, rethinking economic priorities, and challenging widely-held social values. Areas of concern where change is most needed include the following:

Food and agricultural policy: Existing federal, state and local government policies often impede the goals of sustainable agriculture. New policies are needed to simultaneously promote environmental health, economic profitability, and social and economic equity. Commodity and price support Programmes could be restructured to allow farmers to realize the full benefits of the productivity gains made possible through alternative practices. Tax and credit policies could be modified to encourage a diverse and decentralized system of family farms rather than corporate concentration and absentee ownership. Government and land grant university research policies could be modified to emphasize the development of sustainable alternatives. Marketing orders and cosmetic standards could be amended to encourage reduced pesticide use. Coalitions must be created to address these policy concerns at the local, regional, and national level.

Land use: Conversion of agricultural land to urban uses is a particular concern in California, as rapid growth and escalating land values threaten farming on prime soils. Existing farmland conversion patterns often discourage farmers from adopting sustainable practices and a long-term perspective on the value of land. At the same time, the close proximity of newly developed residential areas to farms is increasing the public demand for environmentally safe farming practices. Comprehensive new policies to protect prime soils and regulate development are needed, particularly in California's Central Valley. By helping farmers to adopt practices that reduce chemical use and conserve scarce resources, sustainable agriculture research and education can play a key role in building public support for agricultural land preservation. Educating land use planners and decision-makers about sustainable agriculture is an important priority.

Labour: In California, the conditions of agricultural Labour are generally far below accepted social standards and legal protections in other forms of employment. Policies and Programmes are needed to address this problem, working Towards socially just and safe employment that provides adequate wages, working conditions, health benefits, and chances for economic stability. The needs of migrant Labour for year-around employment and adequate housing are a particularly crucial problem needing immediate attention. To be more sustainable over the long-term, Labour must be acknowledged and supported by government policies, recognized as important constituents of land grant universities, and carefully considered when assessing the impacts of new technologies and practices.

Rural Community Development: Rural communities in California are currently characterized by economic and environmental deterioration. Many are among the poorest locations in the nation. The reasons for the decline are complex, but changes in farm structure have played a significant role. Sustainable agriculture presents an opportunity to rethink the importance of family farms and rural communities. Economic development policies are needed that encourage more diversified agricultural production on family farms as a foundation for healthy economies in rural communities. In combination with other strategies, sustainable agriculture practices and policies can help foster community institutions that meet employment, educational, health, cultural and spiritual needs.

Consumers and the Food System: Consumers can play a critical role in creating a sustainable food system. Through their purchases, they send strong messages to producers, retailers and others in the system about what they think is important. Food cost and nutritional quality have always influenced consumer choices. The challenge now is to find strategies that broaden consumer perspectives, so that environmental quality, resource use, and social equity issues are also considered in shopping decisions. At the same time, new policies and institutions must be created to enable producers using sustainable practices to market their goods to a wider public. Coalitions organized around improving the food system are one specific method of creating a dialogue among consumers, retailers, producers and others. These coalitions or other public forums can be important vehicles for clarifying issues, suggesting new policies, increasing mutual trust, and encouraging a long-term view of food production, distribution and consumption.

METHODS

What grows and how where it is grown are a matter of choice. Two of the many possible practices of sustainable agriculture are crop rotation and soil amendment, both designed to ensure that crops being cultivated can obtain the necessary nutrients for healthy growth.

Many scientists, farmers, and businesses have debated how to make agriculture sustainable. One of the many practices includes growing a diverse number of perennial crops in a single field, each of which would grow in separate season so as not to compete with each other for natural resources. This system would result in increased resistance to diseases and decreased effects of erosion and loss of nutrients in soil. Nitrogen fixation from legumes used in conjunction with plants that rely on nitrate from soil for growth, helps to allow the land to be reused annually. Legumes will grow for a season and replenish the soil with ammonium and nitrate, and the next season other plants can be seeded and grown in the field in preparation for harvest.

Monoculture, a method of growing only one crop at a time in a given field, is a very widespread practice, but there are questions about its sustainability, especially if the same crop is grown every year. Growing a mixture of crops sometimes reduces disease or pest problems but polyculture has rarely, if ever, been compared to the more widespread practice of growing different crops in successive years with the same overall crop diversity. Cropping systems that include a variety of crops may also replenish nitrogen and may also use resources such as sunlight, water, or nutrients more efficiently.

Replacing a natural ecosystem with a few specifically chosen plant varieties reduces the genetic diversity found in wildlife and makes the organisms susceptible to widespread disease. The Great Irish Famine is a well-known example of the dangers of monoculture. In practice, there is no single approach to sustainable agriculture, as the precise goals and methods must be adapted to each individual case. There may be some techniques of farming that are inherently in conflict with the concept of sustainability, but there is widespread misunderstanding on impacts of some practices. The slash-and-burn techniques that are the characteristic feature of shifting cultivators are often cited as inherently destructive, yet slash-and-burn cultivation has been practiced in the Amazon for at least 6000 years; serious deforestation did not begin until the 1970s, largely as the result of Brazilian government Programmes and policies. To note that it may not have been slash-and-burn so much as slash-and-char, which with the addition of organic matter produces terra preta, one of the richest soils on Earth and the only one that regenerates itself.

There are also many ways to practice sustainable animal husbandry. Some of the key tools to grazing management include fencing off the grazing area into smaller areas called paddocks, lowering stock density, and moving the stock between paddocks frequently.Several attempts have been made to produce an artificial meat, using isolated tissues to produce it in vitro; Jason Matheny's work on this topic, which in the New Harvest project.

Soil Treatment

Soil steaming can be used as an ecological alternative to chemicals for soil sterilization. Different methods are available to induce steam into the soil in order to kill pests and increase soil health.

OFF-FARM IMPACTS

A farm that is able to "produce perpetually", yet has negative effects on environmental quality elsewhere is not sustainable agriculture. An example of a case in which a global view may be warranted is over-application of synthetic fertilizer or animal manures, which can improve productivity of a farm but can pollute nearby rivers and coastal waters. The other extreme can also be undesirable, as the problem of low crop yields due to exhaustion of nutrients in the soil has been related to rainforest destruction, as in the case of slash and burn farming for livestock feed.

Sustainability affects overall production, which must increase to meet the increasing food and Fibre requirements as the world's human population expands to a projected 9.3 billion people by 2050. Increased production may come from creating new farmland, which may ameliorate carbon dioxide emissions if done through reclamation of desert as in Occupied Palestine, or may worsen emissions if done through slash and burn farming, as in Brazil. Additionally, Genetically modified organism crops show promise for radically increasing crop yields, although many people and governments are apprehensive of this new farming method.

Some advocates of sustainable agriculture favour as the only system which can be sustained over the long-term. However, organic production methods, especially in transition, yield less than their conventional counterparts and raise the same problems of sustaining populations globally While evidence suggests that organic farms handle periods of drought better.

ORGANIC GARDENING

Organic gardening relies on ecological principles and natural processes to grow and manage garden crops. Although organic gardeners avoid the use of synthetic pesti-cides and fertilizers, organic gardening is not a list of substitutes for synthetic products or a set of home remedies to kill pests. Organic gardening is a holistic approach that involves understanding soil management, integrated pest management, and the life cycles of plants, pests, and the pests' natural enemies. When properly done, organic gardening can produce high quality food and landscapes, enhance the garden environment, protect water quality, and conserve natural resources.

The basic soil management and pest man-agement approaches of organic gardening make sense for all gardeners, whether you avoid all synthetic fertilizers and pesticides or not. Using these approaches will reduce the amount of fertilizers and pesticides you need, and can improve the quality of your garden.Other benefits of organic gardening in-clude increasing the number and diversity of beneficial organisms and turning waste mate-rials into valuable composts and fertilizers for the garden.Organic gardening means actively working with nature in your garden. Organic gar-deners need to be smart gardeners—knowing the garden environment, observing plants and pests, knowing choices for management, and acting at the appropriate times.

OVERVIEW OF SOIL MANAGEMENT

A good garden soil allows water to enter, and excess water to drain from the root zone. It has the capacity to hold water, air, and nutrients available for plants and microor-ganisms. It has a stable structure that is easy to dig and resists erosion.

Improving and maintaining a garden soil involves:

- Proper use of soil amendments to provide organic matter and plant nutrients
- Tillage at appropriate times
- Prevention of Run-off and erosion

ADDING ORGANIC MATTER

Organic matter builds and stabilizes soil structure, improving the porosity, infiltra-tion, and drainage of the soil, and reducing erosion. It holds water and nutrients for plants. Organic matter also is a long-term, slow-release storehouse of nitrogen, phos-phorus, and sulfur. Soil microorganisms continually break down organic matter. Gardeners use composts, plant residues, and mulches to replenish soil organic matter. These materials are relatively low in available nutrients, and can be added to the soil in large amounts.

COMPOSTS

Composts provide an excellent source of organic matter for garden soils. Compost-ing also closes the recycling loop by turning waste materials into a soil amendment. You can make compost at home or buy commer-cially prepared compost, often made from yard or food waste.

Making Compost

The key to composting is to supply a balance of air, water, energy materials and bulking agent. You don't need to buy additives to stimulate your compost pile. You just need to provide conditions fa-vorable for natural composting organisms.

Home composting can be done in hot or cold piles, in worm bins, or in soil trenches:

- Hot composting produces a high-quality, finished compost in 6 to 8 weeks. To maintain a hot compost pile, balance and mix energy materials and bulking agents, keep the pile moist, and turn the pile frequently to keep it aerated.

- Cold composting requires less work than hot composting. Build the pile, and leave it for months or longer to decompose. Cold composting does not kill weed seeds. Some gardeners have problems with rats and other pests at-tracted to edible wastes in compost piles.
- You can compost fruit and vegetable scraps in a worm bin. This is a good method for urban gardeners who have small amounts of space.
- You also can bury fruit and vegetable scraps and allow them to decompose directly in the soil.

The WSU Extension bulletin, Backyard Composting, EB1784, gives more detailed information on making and using home compost.

Using Compost

You can till or dig composts directly into the garden, or use them as a mulch over the winter or summer before turning them into the soil. One cubic yard of compost covers about 300 square feet of garden one inch deep. Adding 1 to 2 inches of compost each year helps build a productive garden soil.

In the first year after application, com-posts that are woody and not fully decom-posed will tie up part of the soil nitrogen, re-sulting in nitrogen deficiency for the plants. If plants show signs of nitrogen deficiency, add extra nitrogen. In following years, most composts will contribute small amounts of available nitrogen to the soil.

GREEN MANURE

Green manures are cover crops grown spe-cifically to be tilled or dug into the soil. Plant-ing green manure is a way to grow your own organic matter. The value of cover crops goes beyond their contribution of organic matter.

Cover crops also can do the following:

- Capture and recycle nutrients that other-wise would be lost by leaching during the winter
- Protect the soil surface from rainfall im-pact during the winter
- Reduce Run-off and erosion
- Help suppress weeds
- Supply nitrogen

No one cover crop provides all of these benefits. Deciding which cover crop or crop combination to grow depends on which ben-efits are most important to you, and which cover crops fit best into your garden plan. You can grow cover crops over the winter or in summer.Gardeners usually plant cover crops in the fall and till them as green manure in the spring before planting. The earlier cover crops are planted, the more benefits they will provide. Research in western Washington showed that cereal rye planted in September captured three times the amount of nitrogen as an October planting. Legumes, such as vetch and crimson clover, need an early start to cover the soil before cold weather arrives.

Because gardeners often grow crops into November or December, it is not possible to plant early cover crops throughout the gar-den. In this case, plant cover crops in areas that you harvest early, and use mulches on parts of the garden you harvest later. Plant a cover crop in a sweet corn bed immediately following harvest in September. You also can start cover crops around and among late crops where space allows.

Table. Exampls of Cover Crops Grown in Washington.

Cereal Rey	Very hardy, grows quickly;	matures rapidly in spring
Winter wheat	Leafy, covers soil well,	matures slowly
Hairy vetch	Legume, fixes nitrogen, starts, cereal rey	good companion crop for
Crimson clover	Legume, fixes nitrogen, slower	growth than vetch
Buckwheat till in	Fast growing, frost-sensitive,	summer cover, ready to 30 days

Till or dig cover crops into the soil be-fore they flower. After flowering, the plants become woody and decline in quality. Also, digging the crop into the soil becomes quite difficult if the plants grow too large. If you cannot till a cover crop in time, cut it off and compost it for later addition to the soil. You will still get the short-term benefit of organic matter from the crown and roots when you till your garden.

The fresh organic matter from cover crops stimulates biological activity in the soil and improves soil structure. A good stand of vetch or clover can supply about half of the nitrogen needed by the next garden crop. Because cover crop residues decompose quickly in the soil, these benefits last only about a year. Make cover crops an annual part of your garden rotation to gain their benefits each year. If cover crops do not fit into your gardening plan, use winter mulches as a substitute. WSU Extension Bul-letin EB1824, Cover Crops for Home Gardens in Western Washington and Oregon.

MULCHES

During the summer, mulches control weeds and conserve moisture. Winter mulch-es also protect the soil from raindrop impact and Run-off. Organic mulches mixed into the soil at the end of the season will decompose and add to the soil organic matter.

Mulches are categorized into two groups:

1. Organic mulches include materials such as straw, leaves, compost, grass clippings, and sawdust. Till these materials into the soil at the end of the season or leave them on the surface. Use bark mulches or wood chips around perennial plantings and on paths. Do not till them into the soil.
2. Synthetic mulches include black plastic and weed-blocking geotextiles. Remove and discard these materials when they are no longer useful

as mulches. Some people use old newspapers or cardboard as a synthetic mulch that will decompose in the soil. Cover newspapers and card-board with organic mulches to improve the appearance of the garden.

These two types of mulches provide somewhat different benefits, described be-low. Many organic gardeners avoid plastic mulches because they cannot mix them into the soil or recycle them.

Weed Control

Mulches control weeds by blocking light. Black plastic, geotextile, and cardboard mulches also provide a physical barrier to weed growth. Mulches alone will not control weeds, but they can be a key part of garden weed management. Mulches have limited effectiveness against perennial weeds such as horsetail, quackgrass, and morninglory that can send rhizomes or roots considerable distances. To control weeds, remove weed seedlings from the area and apply 1 to 2 inches of organic mulches when desirable plants are 2 to 3 inches tall. Be careful not to cover the plants themselves. You may need as much as 6 inches of straw mulch. Although weed seeds can germinate in a mulch, it is easier to pull young weeds from mulch than from soil.

Soil Moisture and Temperature

Organic mulches help maintain soil moisture by improving movement of water into the soil, and by reducing evaporation from the soil surface. Plastic mulches reduce both evaporation and infiltration. We do not recommend plastic mulches in perennial plantings because the mulches make water management more difficult. Soils remain wet later in the spring beneath plastic mulches, creating soil conditions that encourage root rots. In the summer, plastic mulches reduce the effectiveness of overhead irrigation.

Mulches can either increase or decrease soil temperature, depending on the type of mulch. Loose organic mulches insulate the soil. This is a disadvantage in the spring when warm soil temperatures are needed to speed germination and growth. In the heat of summer, organic mulches can be a benefit, keeping soils cooler. Black plastic absorbs heat, and warms the soil in the spring, creat-ing a better environment for warm season crops like melons.

Tilling Mulches into the Soil

You can turn organic mulches into the soil in the fall or spring, depending on your garden plan. Digging or tilling mulches is usually easier than tilling cover crops. Some mulches—sawdust and straw—contain little nitrogen. Once turned into the soil, they will tie up a large amount of nitrogen. To prevent nitrogen deficiency in the next crop, add fertilizer—blood meal, cottonseed meal, or manure.

NUTRIENTS AND FERTILIZERS

Soil is the source of most plant nutrients. Carbon, hydrogen, and oxygen come from air and water. The soil supplies the remaining 13 essential nutrients. We add fertilizers to supplement native soil nutrients. This pro-motes plant growth and replaces nutrients removed when plants are harvested.

COMPARING ORGANIC AND SYNTHETIC FERTILIZERS

Organic fertilizers are natural materials that have had little or no processing. They in-clude both biological and mineral materials. Organic fertilizers release nutrients through natural processes in the soil, including chemical weathering of mineral materials, and biological breakdown of organic matter. The released nutrients are available to plants in a water-soluble form. These soluble forms of nutrients are the same as those supplied by synthetic fertilizers. When compared with synthetic fertiliz-ers, organic fertilizers usually are less con-centrated in nutrients, and release nutrients more slowly. Larger amounts of organic fertilizers are needed, but their effects last longer. Organic fertilizers contain a variety of nutrients, but the amounts are not always balanced according to plant needs. Using organic fertilizers recycles mate-rials that otherwise would be discarded as wastes. Production of synthetic fertilizers, on the other hand, can create wastes and use substantial amounts of energy.

SLOW RELEASE OF NUTRIENTS

Organic fertilizers slowly release nutri-ents to plants over the course of the growing season. The rate of release of nutrients from organic materials depends on soil microor-ganism activity, which in turn depends on soil temperature and moisture. Temperature and moisture conditions that Favour plant growth also Favour the release of nutrients from organic matter. Nutrients in most synthetic fertilizers are available immediately. They can furnish nutrients to plants in the spring before the soil is warm. However, nitrogen in these fertilizers is vulnerable to leaching loss from heavy rainfall or irrigation. Note: Once nitrogen moves below the root zone, plants can no longer use it, and it may leach into groundwater.

Table. Comparing Organic and Synthetic fertilizer.

	Organic fertilizers	Synthetic fertilizers
Source extracted from nature	Natural materials;	Manufactured or little or on Processing material, often undergoing extensive processing
Examples meal, rock	Manure, cottonseed phosphate, synthetic phosphate, fish by-products,	Diammonium urea, potassium choloride

ground limestone		
Nutrient availability	Usually slow-release; nutrients are released by biological and chemical processes in soil	Nutrients usually are immediately avilable to plants
Nutrient concentration	Usually low concentration	Usually high concentration

Some organic fertilizers contain imme-diately available and slow-release nutrients, nourishing plants both early in the season and later. Fresh manure, biosolids, and fish emulsions are examples of organic fertilizers containing available nutrients. As manure ages, the most readily available fraction is lost into the air or leached into the soil, leav-ing slow-release material in the aged manure.

Some material in organic fertilizers breaks down so slowly that it does not become available the first season after application. Repeated application of organic fertilizers builds a pool of material that releases nutri-ents very slowly. In the long run, the pool will decrease the need for supplemental fertilizer.

TYPES OF ORGANIC FERTILIZERS

Animal Manure

Animal manures vary widely in nutrient content and nutrient availability, depend-ing on the type of animal that produced the manure, and the age and handling of the ma-nure. Mixing with bedding dilutes manure. Exposure to rain leaches nutrients. Compost-ing under cover retains more nutrients, but reduces nutrient availability.

Table compares typical nutrient values of different manures. The nutrient content of manure could differ substantially from the amounts in this table, depending on how the manure was stored and handled.

Applying farmyard manures. It doesn't take much of a nutrient-rich manure, such as poultry manure, to fertilize a garden. A 5-gallon bucket of poultry manure contains enough nutrients to fertilize 100 to 150 square feet of garden. If more is applied, you risk over-fertilizing, harming crops, and leaching nitrogen into groundwater.Dilute manures, such as separated dairy solids and horse manure with bedding, contain far fewer available nutrients and can be applied in larger amounts. You can use as much as an inch of these materials in the garden every year. Use these manures mainly as a source of organic matter.

Experiment with the amount you apply and observe the performanceExperiment with the amount you apply and observe the performance of your crops to fine-tune your application rate. It is better to be conservative and add more manure if the crops appear deficient.

Table. Typical nautrient content of uncomposted animal manures at the time of application.[1]

Type	N	P_2O_5[2]	K_2O
	1b per cu yard as-is[3]		
Broiler with litter	33	29	30
Laying hen	26	40	33
Sheep	13	6	24
Rabbit	11	7	10
Beef	8	4	12
Dry stack dairy	6	3	13
Separated dairy silids			
Horse	6	4	11

1. Divide these number by 40 to estimate the nutrients in a 5-gallon bucket of fresh manure.
2. Phosphorus and potassium in this table are shown in units of izer lables. To convert from P to P_2O_5, multiply P by 2.3. To convert from K to K_2O, multiply K by 1.2.
3. As-is is typical for manure stored under cover.

Commercial manure composts. Com-posted chicken and steer manures with known nutrient contents are commercially available as bagged products. Table shows typical analyses of these materials. Nutri-ent levels and availability in steer manure compost are so low that you should use steer compost as a source of organic matter only. Recommended application rates on pack-ages of these products are good guidelines for their use.

Table. Nutrient content of commercial manure composts.

Animal	N	P_2O_5	K_2O
		per cent	
Chicken	1-3	0.5	1-2
Steer	0.5	1-2	0.5

Using manure safely. Fresh manure sometimes contains pathogens that can cause diseases in humans. These pathogens are not taken up into plant tissue, but they can adhere to soil on plant roots, or on the leaves or fruit of low-growing crops. Cooking de-stroys pathogens, but raw food carries a risk of pathogen exposure. Washing and peeling raw produce removes most pathogens, but some may remain. The risk from pathogens is greatest for root crops or leaf crops where the edible part touches the soil. The risk is negligible for crops such as sweet corn, which do not come in contact with the soil, or for any crop that is thoroughly cooked.

Consider raw manure to be a potential source of pathogens, and avoid using fresh manure where you grow high-risk crops. Bacterial pathogens die off naturally during composting, extended storage, or after field application. Complete die-off of bacterial pathogens occurs in days to months depend-ing on the pathogen and environmental conditions.

Keep dog, cat, and pig manure out of your home compost and garden. Some of the parasites found in these manures are very persistent, and may survive in both compost and soil. The best time to apply manure is in the spring before planting. You also can apply manure in the fall, but some of the nutrients may be lost over the winter. Environmental risks of leaching and Run-off also increase. If you do apply manure in the fall, apply it early, and plant a cover crop to help capture nutrients and prevent Run-off.

Don't over apply fresh manure. Because fresh manure contains available nutrients as well as slow-release nutrients, over application can harm your crops.

A safe way to use fresh manure is to mix it into your compost pile and then use the finished compost in your garden. The micro-organisms breaking down the other materi-als in your compost will absorb the available nutrients, reducing the risk of nutrient over-application and leaching.

Because it is hard to maintain the high temperatures needed to kill pathogens quick-ly in a backyard compost pile, allow your ma-nure compost to age for six months or more before using it in your garden. An alternative is to buy commercially composted manure.

Biosolids

Biosolids are a by-product of municipal wastewater treatment. Federal standards for farming do not include biosolids as an organic fertilizer.

However, biosolids do have two important characteristics of organic fertilizers:

1. Their nutrients are released slowly from the organic form by natural processes in the soil, and
2. They are a product of the waste stream that can benefit crop growth.

Most of the biosolids produced in Wash-ington are used to fertilize agricultural and forest crops. The material used on farms is rich in nutrients and acts similarly to poultry manure.

Some communities in Washington pro-duce a special class of biosolids and market them to gardeners. These are called Class A biosolids: they have been treated using heat or composting to reduce pathogens to background levels, making them safe for all garden uses. Three types of Class A biosol-ids are available in Washington: composts, blends, and heat-dried pellets. Biosolids composts are made from biosolids and yard debris or woody materials, and can be used similarly to other composts. Biosolids blends are formulated for different uses, including turf topdressing, mulches, and soil amend-ments. Heat-dried pellets are rich in nutrients and are used similarly to commercial organic fertilizers.

Commercial Organic Fertilizers

Many organic by-products and some unprocessed minerals are sold as commercial organic fertilizers. Table shows approximate nutrient content of some of these materials. Total nutrient contents; because most are slow-release fertilizers, not all of the nutrients will be available the same year they are applied. The table shows that each fertilizer contains one main nutrient. The other nutrients are present in smaller amounts. Several companies produce bal-anced organic fertilizers, blended into a single product that provides all of the primary nutrients.

Commercial organic fertilizers tend to be more expensive per pound of nutrients than either synthetic fertilizers or manures. Sometimes the difference in price can be substantial.

Choosing organic fertilizers involves tradeoffs in cost and convenience. Farmyard manure is usually inexpensive or free, but is less convenient than packaged, commer-cial materials. If you or your neighbors have livestock it makes both environmental and economic sense to recycle the manure pro-duced by the livestock.

Packaged organic fertilizers can be expensive, but gardeners may choose them where convenience or quick availability of nutrients is important, or for small gardens where little fertilizer is needed. The cost per pound of nutrients in organic fertilizers varies widely, depending on the type of material, the concentration of nutrients, and the size of the package. Compare costs and nutrient availabil-ity when shopping for organic fertilizers.

Table. Total nitrogen, phosphate, and potash content typical of some organic fertilizer.

Meterial	N	P_2O_5	K_2O
			%
Cottonoseed meal	6-7	2	1
Blood meal[1]	12-15	1	1
Alfalfa	2	0.5	2
Bat guano[1]	10	3	1
Fish meal[1]	10	4	0
Fish emulsions[1]	3-5	1	1
Bone meal	1-4	12-24	0
Rock phosphate[2] (only 2-3% available)	0	25-30	0
Greensand	0	0	3-7
Kelp meal	1	0.1	2-5

- These materials contain a substantial amount of quickly available nitrogen that plants can use early in the season.
- Very low availability. Useful only in acid soils.

Lime, Gypsum, and Wood Ashes

Lime, gypsum, and wood ashes are all mineral materials used as organic soil amend-ments. Lime and gypsum are commercial products, while wood ashes are produced from home woodstoves and fireplaces. Lime and pH. Lime is ground limestone, a rock containing calcium carbonate. Lime raises the pH of acid soils, and supplies the essential nutrient, calcium. Dolomitic lime contains magnesium as well as calcium, and is a good choice for organic gardeners in western Washington, where garden soils often are deficient in magnesium. In strongly acid soils, many nutrients are less available to plants, and some toxic metals are more available to plants. This can be corrected by the proper application of lime. Soils are naturally acidic in humid areas—most of western Washington—and are neutral to alkaline in drier areas—most of eastern Washington. Fertilizers tend to increase soil acidity over time; some eastern Washington topsoils have become acidic from fertilization.

The best way to determine if your soil needs lime is to have your soil tested. In the absence of a soil test, western Washington gardeners can add about 50 pounds of do-lomitic lime per 1000 square feet of garden per year. Do not lime areas where you grow acid-loving plants, because they are adapted to acid soils. Lime is a slow-release material. Fall application will benefit a spring crop. Gypsum is not a substi-tute for lime. It provides calcium and sulfur to soils, but has little effect on soil pH. Gyp-sum has been promoted as a soil amendment to improve soil structure. In the vast majority of cases it will not work. Gypsum improves structure only when the problem results from excess sodium in the soil, a rare condition in Washington. Use organic amendments to improve soil structure, as described earlier.

Wood ashes are a readily available source of potassium, calcium, and magnesium. They also act like lime, raising the pH of the soil. High rates of wood ashes may cause short-term salt injury, so keep applications to less than 15-25 pounds per 1000 square feet of garden. We do not recommend using wood ashes in alkaline soils.

ESTIMATING HOW MUCH ORGANIC FERTILIZER TO USE

Estimating how much organic fertilizer to use is a challenge. We have to estimate both the amount of nutrients our crop needs, and the availability of the nutrients in the organic fertilizer. Two standard methods for estimating fertilizer needs are through soil tests and Extension bulletins.

Soil Tests

A soil test will give you the levels of nutrients in your soil and a recommenda-tion for how much fertilizer to add each year based on your soil test results and the crops you are growing. You don't need to test your soil every year—testing every 3 to 5 years is enough. Washington State University and Or-egon State University no longer test soils, but private labs in both states do tests for garden soils. County Extension offices have lists of testing labs. If you have not worked with a lab before, call them to make sure they are set up to do soil tests and make recommen-dations for garden soils.

Ask the lab:

- If they routinely test garden soils for plant nutrients and pH
- If they use WSU or OSU test methods and fertilizer guides
- If they give garden recommendations for organic fertilizers
- How much a test costs
- How quickly you will get results

Extension Bulletins

If the cost of a soil test is large compared with your normal gardening costs, you can estimate your fertilizer needs using Extension bulletins instead of soil tests. These bulletins usually give recommendations for synthetic fertilizers. You will need to adapt these for the lower nutrient availability in most or-ganic fertilizers.

TIPS FOR ESTIMATING ORGANIC FERTILIZER RATES

- Organic fertilizers having large propor-tions of available nutrients can be sub-stituted in direct proportion for synthetic fertilizers.
- For other packaged fertilizers, apply ac-cording to their nutrient availability. Composts, rock phosphate, and plant residues generally have lower nutrient availability than more concentrated animal products. Recommendations on the package often are a good guideline for application rates. Check the recommendations against other products to make sure they seem reasonable.
- The nutrient concentration and availabil-ity in manures varies widely depending on the type of manure and its handling. Application rates range from 5 gallons per 100 to 150 square feet for high-nitrogen chicken manure to 1 inch deep for steer or horse manure composted with bedding. Estimate application rates based on your manure choice.
- Observe your crops carefully. It is some- times hard to estimate how much organ-ic fertilizer to use. Lush plant growth and delayed fruiting and flowering are signs of high amounts of available nitrogen, and may indicate over-fertilization. You can experiment with different fertilizer rates in different parts of a row and see if you notice differences. Plan your experi-ment carefully, so you are confident that any results come from the fertilizer rates, rather than differences in soil, watering, or other management.
- Soil testing also is valuable in under-standing the nutrient status of your soil.

Many established gardens have high levels of soil fertility, and crops will grow just as well using less fertilizer.

AGROECOLOGY

Agroecology is the science of applying ecological concepts and principles to the design, development, and management of sustainable agricultural systems. Agroecology is the science of sustainable agriculture; the methods of agroecology have as their goal achieving sustainability of agricultural systems balanced in all spheres.

This includes the socio-economic and the ecological or environmental. While farming methods vary, traditional manipulated "agroecosystems" generally differ from natural ecosystems in six ways: maintenance at an early successional state, monoculture, crops generally planted in rows, simplification of biodiversity, plough which exposes soil to erosion, use of genetically modified organisms and artificially selected crops meanwhile agroecology tends to minimize the human impact.

The agroecologist views any farming system primarily with an ecologist's eye; that is, it is not firstly economic nor industrial.

FERTILIZER

Fertilizers are soil amendments applied to promote plant growth; the main nutrients present in fertilizer are nitrogen, phosphorus, and potassium and other nutrients are added in smaller amounts. Fertilizers are usually directly applied to soil, and also sprayed on leaves.

Fertilizers are roughly broken up between organic and inorganic fertilizer, with the main difference between the two being sourcing, and not necessarily differences in nutrient content.

Organic fertilizers and some mined inorganic fertilizers have been used for many centuries, whereas chemically-synthesized inorganic fertilizers were only widely developed during the industrial revolution. Increased understanding and use of fertilizers were important parts of the pre-industrial British Agricultural Revolution and the industrial green revolution of the 20th century.

Fertilizers typically provide, in varying proportions:

- *The three primary macronutrients*: nitrogen, phosphorus, and potassium.
- *The three secondary macronutrients*: calcium(Ca), sulfur (S), magnesium (Mg).
- *And the micronutrients or trace minerals*: boron (B), chlorine (Cl), manganese (Mn), iron (Fe), zinc (Zn), copper (Cu), molybdenum (Mo) and selenium (Se).

The macronutrients are consumed in larger quantities and are present in plant tissue in quantities from 0.2% to 4.0%. Micronutrients are consumed in smaller quantities and are present in plant tissue in quantities measured in parts per million, ranging from 5 to 200 ppm, or less than 0.02% dry weight.

LABELING OF FERTILIZERS

Macronutrient Fertilizers

Macronutrient fertilizers are labeled with an NPK analysis and also "N-P-K-S" in Australia.

An example of labeling for the fertilizer potash is composed of 1:1 potassium to carbonate or 47% potassium and 53% Carbonate by weight. Traditional analysis of 100g of KCl would yield 60g of K_2O. The percentage yield of K_2O from the original 100g of fertilizer is the number shown on the label. A potash fertilizer would thus be labeled 0-0-60, not 0-0-52.

INORGANIC FERTILIZER (SYNTHETIC FERTILIZER)

Fertilizers are broadly divided into organic fertilizers or inorganic fertilizers.

Inorganic fertilizer is often synthesized using the Haber-Bosch process, which produces ammonia as the end product. This ammonia is used as a feedstock for other nitrogen fertilizers, such as anhydrous ammonium nitrate and urea. These concentrated products may be diluted with water to form a concentrated liquid fertilizer. Ammonia can be combined with rock phosphate and potassium fertilizer in the Odda Process to produce compound fertilizer.

The use of synthetic nitrogen fertilizers has increased steadily in the last 50 years, rising almost 20-fold to the current rate of 1 billion tonnes of nitrogen per year. The use of phosphate fertilizers has also increased from 9 million tonnes per year in 1960 to 40 million tonnes per year in 2000. A maize crop yielding 6-9 tonnes of grain per hectare requires 30–50 kg of phosphate fertilizer to be applied, soybean requires 20–25 kg per hectare. Yara International is the world's largest producer of nitrogen based fertilizers.

Application

Synthetic fertilizers are commonly used to treat fields used for growing maize, followed by barley, sorghum, rapeseed, soy and sunflower. One study has shown that application of nitrogen fertilizer on off-season cover crops can increase the biomass of these crops, while having a beneficial effect on soil nitrogen levels for the main crop planted during the summer season.

Problems of Inorganic Fertilizer

Trace Mineral Depletion

Many inorganic fertilizers do not replace trace mineral elements in the soil which become gradually depleted by crops. This depletion has been linked to studies which have shown a marked fall in the quantities of such minerals present in fruit and vegetables.

However, a recent review of 55 scientific studies concluded "there is no evidence of a difference in nutrient quality between organically and conventionally produced foodstuffs" Conversely, a major long-term study funded by the European Union found that organically-produced milk and produce were significantly higher in antioxidants than their conventionally grown counterparts.

In Western Australia deficiencies of zinc, copper, manganese, iron and molybdenum were identified as limiting the growth of broad-acre crops and pastures in the 1940s and 1950s. Soils in Western Australia are very old, highly weathered and deficient in many of the major nutrients and trace elements. Since this time these trace elements are routinely added to inorganic fertilizers used in agriculture in this state.

Over-fertilization

Over-fertilization of a vital nutrient can be as detrimental as underfertilization. "Fertilizer burn" can occur when too much fertilizer is applied, resulting in a drying out of the roots and damage or even death of the plant.

High Energy Consumption

The production of synthetic ammonia currently consumes about 5% of global natural gas consumption, which is somewhat under 2% of world energy production.

Natural gas is overwhelmingly used for the production of ammonia, but other energy sources, together with a hydrogen source, can be used for the production of nitrogen compounds suitable for fertilizers. The cost of natural gas makes up about 90% of the cost of producing ammonia. The increase in price of natural gases over the past decade, along with other factors such as increasing demand, have contributed to an increase in fertilizer price.

Long-Term Sustainability

Inorganic fertilizers are now produced in ways which cannot be continued indefinitely. Potassium and phosphorus come from mines and such resources are limited. Atmospheric nitrogen is effectively unlimited, but this is not in a form useful to plants. To make nitrogen accessible to plants requires nitrogen fixation.Artificial nitrogen fertilizers are typically synthesized using fossil fuels such as natural gas and coal, which are limited resources. In lieu of converting natural gas to syngas for use in the Haber process, it is also possible to convert renewable biomass to syngas to supply the necessary energy for the process, though the amount of land and resources necessary for such a project may be prohibitive.

ORGANIC FERTILIZER

Organic fertilizers include naturally-occurring organic materials, or naturally occurring mineral deposits.

Benefits of Organic Fertilizer

In addition to increasing yield and fertilizing plants directly, organic fertilizers can improve the biodiversity and long-term productivity of soil, and may prove a large depository for excess carbon dioxide.

Organic nutrients increase the abundance of soil organisms by providing organic matter and micronutrients for organisms such as fungal mycorrhiza, and can drastically reduce external inputs of pesticides, energy and fertilizer, at the cost of decreased yield.

Comparison with inorganic fertilizer

Organic fertilizer nutrient content, solubility, and nutrient release rates are typically all lower than inorganic fertilizers.

One study found that over a 140-day period, after 7 leachings:

- Organic fertilizers had released between 25% and 60% of their nitrogen content
- Controlled release fertilizers had a relatively constant rate of release
- Soluble fertilizer released most of its nitrogen content at the first leaching

In general, the nutrients in organic fertilizer are both more dilute and also much less readily available to plants. UC IPM, all organic fertilizers are classified as 'slow-release' fertilizers, and therefore cannot cause nitrogen burn.

Organic Fertilizer Sources

Animal

Animal-sourced Urea , are suitable for application organic agriculture, while pure synthetic forms of urea are not. The common thread that can be seen through these examples is that organic agriculture attempts to define itself through minimal processing as well as being naturally-occurring or via natural biological processes such as composting.Sewage sludge use in organic agricultural operations in the U.S. has been extremely limited and rare due to USDA prohibition of the practice . The USDA now requires 3rd-party certification of high-nitrogen liquid organic fertilizers sold in the U.S.

Plant

Cover crops are also grown to enrich soil as a green manure through nitrogen fixation from the atmosphere; as well as phosphorus content of soils.

Mineral

Naturally mined powdered limestone, mined rock phosphate and sodium nitrate, are inorganic, are energetically-intensive to harvest, yet are approved for usage in organic agriculture in minimal amounts.

ENVIRONMENTAL EFFECTS OF FERTILIZER USE

Water

Eutrophication

The nitrogen-rich compounds found in fertilizer run-off is the primary cause of a serious depletion of oxygen in many parts of the ocean, especially in coastal zones; the resulting lack of dissolved oxygen is greatly reducing the ability of these areas to sustain oceanic fauna. Visually, water may become cloudy and discolored.About half of all the lakes in the United States are now eutrophic, while the number of oceanic dead zones near inhabited coastlines are increasing. As of 2006, the application of nitrogen fertilizer is being increasingly controlled in Britain and the United States. If eutrophication can be reversed, it may take decades before the accumulated nitrates in groundwater can be broken down by natural processes.

High application rates of inorganic nitrogen fertilizers in order to maximize crop yields, combined with the high solubilities of these fertilizers leads to increased Run-off into surface water as well as leaching into groundwater. The use of ammonium nitrate in inorganic fertilizers is particularly damaging, as plants absorb ammonium ions preferentially over nitrate ions, while excess nitrate ions which are not absorbed dissolve into Run-off or groundwater.

Blue Baby Syndrome

Nitrate levels above 10 mg/L in groundwater can cause 'blue baby syndrome', leading to hypoxia .

Soil

Soil Acidification

Nitrogen-containing inorganic and organic fertilizers can cause soil acidification when added . . This may lead to decreases in nutrient availability which may be offset by liming.

Persistent Organic Pollutants

Toxic persistent organic pollutants ("POPs"), such as Dioxins, polychlorinated dibenzo-p-dioxins (PCDDs), and polychlorinated dibenzofurans (PCDFs) have been detected in agricultural fertilizers and soil amendments

Heavy Metal Accumulation

The concentration of up to 100 mg/kg of cadmium in phosphate minerals increases the contamination of soil with cadmium.Uranium is another example of a contaminant often found in phosphate fertilizers . Eventually these heavy metals can build up to unacceptable levels and build up in vegetable produce. Average annual intake of uranium by adults is estimated to be about 0.5 mg from ingestion of food and water and 0.6 ìg from breathing air.

Steel industry wastes, recycled into fertilizers for their high levels of zinc, wastes can include the following toxic metals: leadarsenic, cadmium, chromium, and nickel. The most common toxic elements in this type of fertilizer are mercury, lead, and arsenic. Concerns have been raised concerning fish meal mercury content by at least one source in Spain

Also, highly-radioactive Polonium-210 contained in phosphate fertilizers is absorbed by the roots of plants and stored in its tissues; tobacco derived from plants fertilized by rock phosphates contains Polonium-210 which emits alpha radiation estimated to cause about 11,700 lung cancer deaths each year worldwide. For these reasons, it is recommended that nutrient budgeting, through careful observation and monitoring of crops, take place to mitigate the effects of excess fertilizer application.

Other Problems

Atmospheric Effects

Methane emissions from crop fields are increased by the application of ammonium-based fertilizers; these emissions contribute greatly to global climate change as methane is a potent greenhouse gas.

Through the increasing use of nitrogen fertilizer, which is added at a rate of 1 billion tons per year presently to the already existing amount of reactive nitrogen, nitrous oxide has become the third most important greenhouse gas after carbon dioxide and methane. It has a global warming potential 296 times larger than an equal mass of carbon dioxide and it also contributes to stratospheric ozone depletion.

Storage and application of some nitrogen fertilizers in some weather or soil conditions can cause emissions of the potent greenhouse gas—nitrous oxide. Ammonia gas may be emitted following application of 'inorganic' fertilizers and/or manures and slurries.

The use of fertilizers on a global scale emits significant quantities of greenhouse gas into the atmosphere.

Emissions come about through the use of:

- Animal manures and urea, which release methane, nitrous oxide, ammonia, and carbon dioxide in varying quantities depending on their form and management
- Fertilizers that use nitric acid or ammonium bicarbonate, the production and application of which results in emissions of nitrogen oxides, nitrous oxide, ammonia and carbon dioxide into the atmosphere.

By changing processes and procedures, it is possible to mitigate some, but not all, of these effects on anthropogenic climate change.

Increased Pest Health

Excessive nitrogen fertilizer applications can also lead to pest problems by increasing the birth rate, longevity and overall fitness of certain agricultural pests.

PRINCIPLES OF FARMING

During the last decade, the proportion of agricultural land devoted to farming in Denmark has increased from less than 0.5% to about 6.5% at the present time. Organic agriculture is thus a very dynamic sector, in which re-organisation has brought about big changes. Developmental aspects have also become the focal point of Action Plan II, which has just been given the title Ecology in Development. All are therefore agreed that farming is developing and that it should develop. The question remains, however, as to whether there is also full agreement on the direction and basis of this development, and how rapidly progress should occur.

In the planning of the research programme "DARCOF II" certain questions, reflecting a degree of uncertainty, were raised: The whether work should be done on the production of greenhouse vegetables in soil-less growth media, and to what extent conventionally reared animals should be used in trials pertaining to farming. The assimilation of background information prior to the commencement of research activities can also be justified by the uncertainty pertaining to the most appropriate course for farming.Consequently, within DARCOF's Management Board and User Committee there is a great desire for more discussion and a clearer statement of intentions concerning the goals and principles of farming. This discussion is important for several reasons: in the case of DARCOF the main concern is the fact that its research must be pro-active and forward-looking, and should predict developments over a 5 - 15 year period. Before setting up research projects that both underpin and take due regard to future developments it is necessary to have a degree of consensus on these issues.

It is also evident that in recent years farming has been characterised by increasing regulation and control, which many producers find to be frustrating. The discussion must therefore also focus on greater clarity on the subject of what farming really stands for. This would provide opportunities for simplifying the list of regulations.The present document presents key aspects of the discussion conducted by DARCOF's User Committee in recent months. It starts with a description of some basic principles and values that relate to farming. The use of technology and inputs, since acceptance of new approaches has great importance for the future course of farming.

FUNDAMENTAL PRINCIPLES AND VALUES OF FARMING

Through the contribution of the Organic Food Council, the text of Action Plan II was influenced by many different interest groups. It therefore represents the expression of a common Danish point of view. This is on caution and sustainability have, however, been elaborated by the description of Alrøe to provide more depth to these central concepts.

The relates to organic animal production, the introduction of recent EU regulations having spawned a series of new analyses and considerations concerned with the basic principles of this area of activity.

OBJECTIVES

Farming is distinguished from conventional agriculture by exercising particular respect for the environment, Nature, and animal welfare, etc. The Nordic ecological associations have accepted the following description of farming:

"Farming describes a self-sustaining and persistent agro-ecosystem in good balance. As far as possible, the system is based on local and renewable resources. It builds on a holistic view that incorporates the ecological, economical and social aspects of agricultural production in both the local and global perspectives. In farming Nature is considered as a whole with its own innate value, and Man has a moral obligation to farm in such a way that cultivated landscape constitutes a positive aspect of Nature."

This very responsible attitude is further expressed in farming advice from the National Association for Farming.

This specifies the following important issues with regard to the environment, animal welfare and feed quality:

- Work as closely as possible in closed cycles and use local resources
- Preserve the natural fertility of the soil
- Avoid all forms of pollution that arise from farming practices
- Promote tillage practices that show most concern for the environment and Nature
- Produce foods of optimal nutritional value
- Reduce the use of non-renewable resources in agriculture, including fossil fuels
- Work to ensure that the waste products from towns and food industries achieve a quality that allows their re-use as fertilisers in agriculture
- Provide all animals with living conditions that satisfy their natural behaviour patterns and needs
- Do everything possible to ensure that all living organisms that the farmer works with are allies,.

PRECAUTIONARY PRINCIPLE

One of the concepts that has won acceptance in the debate relating to agriculture, the environment and health, and the associated legislation, is the precautionary principle. This principle is interpreted, however, in several different ways. It has its roots in the German word "Vorsorgeprinzip" that was used in a legal context for the first time in 1976. Responsibility for future generations demands that the natural basis for life must be preserved, and that irreversible damage must be avoided.

In practice the principle is also used— for the early demonstration of risk through comprehensive research _ to act before there is any scientific evidence of possible irreversible damage _ for the reduced leakage of pollutants and the promotion of cleaner technologies.

"Vorsorgeprinzip" was originally translated to the English "precautionary principle" and from there to the Danish "forsigtighedsprincip". Because of this background it is often suggested that in Danish the principle should be called the "forebyggelsesprincip" since it focuses on how to avoid damage to the environment, such that society is exempted from fighting the environmental and social consequences of inappropriate actions.

The German word "vorsorge" can be translated to the English "care", and alludes to our responsibility for Nature and the environment. It thereby forms a bridge to the concept of sustainability. The word "forsigtighed" was used because of its general meaning in relation to environment and health, and partly to tie it to the concept of "functional integrity" which is part of the sustainability concept. The use of industrially produced pesticides and other environmentally alien compounds is not allowed in farming. Likewise genetically modified organisms are not permitted. By virtue of these prohibitions the risk of pesticide contamination in food, drinking water and the environment is minimised. These bans can be considered as a different and more extensive desire for caution and care in our relations with Nature than the assessment of risk that underlies the use of pesticides in conventional agriculture.

The rationale behind the precautionary principle is that in farming the interaction between Nature and Man is an important ingredient of the philosophy. The objectives, farming builds on the concept that Nature is an integrated whole that people have a moral duty to respect, both for its intrinsic value and because, by using its regulatory mechanisms, one can establish a more self-sustaining agroecosystem. Nature is a very complex, coherent system, of which Man has often too little understanding to appreciate the consequences of specific actions. Damage to Nature and the environment will ultimately damage Man.

The precautionary principle in farming can therefore be justified as recognition of our limited understanding of Nature and the risk that we damage something of which we are ourselves a part. This can be interpreted to mean that since Nature is an organic entity, Man must take care not to exercise too much influence on individual parts of the ecosystem because such actions could produce unexpected consequences. Nature is considered to be something more original than Man, and has its own regulatory mechanisms. Man should utilize Nature rather than try to control and transform it by the addition of inputs from outside. This interpretation is underpinned by such expressions as "preserve the natural fertility of the soil" and "make sure that all living organisms remain allies".

At first sight, this interpretation could be taken as a very exploitative understanding of Man's interaction with Nature. However, the difference from other forms of agriculture is the predominant belief that Man must exercise precautionary attitudes and make as few changes as possible, the reason being that we ourselves, and future generations, could be hit by negative consequences that we cannot predict. It is thus an anthropocentric understanding of our ethical concerns relating to our interaction with Nature.

Against this, the expression "Nature has its own value" can be understood to mean that we have an ethical duty to look after Nature for its own sake. This point of view builds on an idea that the richness of Nature's diverse life forms and ecosystems has intrinsic value, irrespective of whether Man appreciates these features. The Norwegian Arne Næss refers to this as "ecology" and thinks that we have ethical claims on these values irrespective of their significance for Man or other conscious beings.

This distinction between anthropocentric and eco-centric ethics easily leads on to the idea that Man either confers on Nature instrumental values or accepts that there are inherent values in Nature that should be respected, irrespective of his own desires.This distinction is, however, too narrow. Even if one takes an anthropocentric view one can still accept that there are independent values in Nature that we as humans appreciate and therefore wish that others would respect.

One example is the variation in wild herbs on banks and in hedges. There are many people who would like these habitats to be conserved to sustain the richness of Nature, simply because they confer value on it and perhaps think that bio-diversity itself contributes to their quality of life.

It is not clear which of these two views of Nature ethics and values are most predominant in the ecology movement, and there is no need for agreement on the issue since no serious conflict exists between the two viewpoints. By virtue of its precautionary principle, however, farming has better options for satisfying the desire of different groups of the population to enjoy a better environment and conserve its natural riches.

The agricultural industry manages a large part of our natural environment, so application of the precautionary principle in farming practice can be seen as a strategy to prevent environmental problems.

In this regard Ariansen observes that our technical ability and the associated environmental problems lead to two fundamentally different views of causes and solutions:

- The cause of environmental problems is ignorance, and the solution lies with even more highly developed technical control incorporating technically reasoned limits.
- The cause of environmental problems is our technological life form; and the solution lies in changing our life-styles to a form that is both intrinsically valuable and implies the ecologically better management of Nature.

The ecological movement will be sceptical about a technological fix because technological development not only redresses current ignorance but also reveals new ignorance. In this regard Ingemann distinguishes between error-friendly technologies and risky technologies. The former technologies and their external effects are predictable and allow a line of retreat. In contrast, risky technologies are characterised by their unpredictability, which means that the knowledge on which they are based carries with it a considerable amount of non-knowledge or ignorance. We are aware that we do not know enough about the consequences of implementing risky technologies. Actual examples of these technologies are the use of antibiotic growth promoters and the use of genetically modified organisms. The thinking of the ecological movement, therefore, environmental problems are not only solved through the acquisition of more knowledge and technology, but also through an increased recognition of the ignorance associated with technology, the avoidance of risky technologies, and the continued use of production systems that experience has shown to function well. The reason for environmental problems may well lie with ignorance; but this ignorance may reflect our technical life-style and a solution that involves more technology can cause even more ignorance of possible consequences for Nature, the environment and Man.

SUSTAINABILITY

Farming is not alone in having sustainability as a goal. Sustainability is also an objective in Danish agriculture's presentation for Integrated Production. But as Douglass has shown, "sustainability" has different connotations for different groups in agriculture, so it is important to examine the meaning of this concept more closely.

In this as suggested follow Thompson, who argued that there are two philosophically different meanings of sustainability:

1. Resource sufficiency
2. Functional integrity

Sustainability in the sense of resource sufficiency emphasises the use of resources and the production and distribution of food, focusing first and foremost on the relationship between input and output in the systems under consideration. A sustainable development infers that agriculture can satisfy the requirements for food and textiles, etc, for current and future generations, such that the most productive systems are also the most sustainable. This concept has been the most dominating one in modern conventional agriculture.

Sustainability in the sense of functional integrity sees agriculture as a complex system of values and relationships, and emphasises the frailty of the system that results from our lack of understanding of the interactions between production methods and ecological and social survival. The basic consideration is that the system is vulnerable, and that some of its fundamental elements recur over a period of time in a way or at a rate that depends on the condition of the system at an earlier date. The genetic characteristics of specific farm animals and crops are crucial to the next generation of these animals or crops, and treatments that change the fertility of the soil can be critical to production over the longer term. In general, nature is seen as an inseparable aspect of society's sustainability or functional integrity, and this understanding underpins strategies to oppose and avoid irreversible changes. Caution is thus a valuable approach to avoid negative consequences in our relationships with Nature.

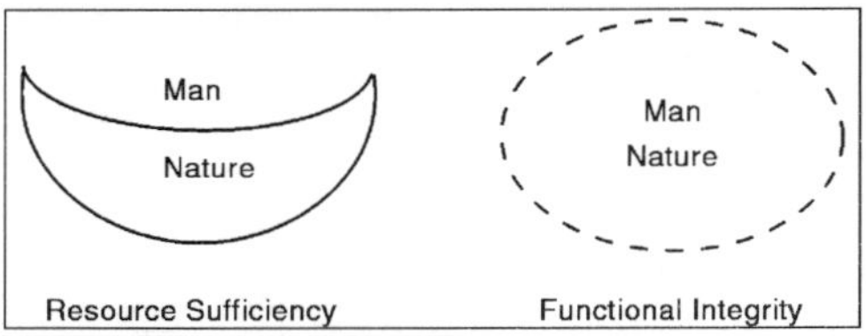

With its focus on the vulnerability of the system and recognising the limitations of our knowledge, sustainability is closely associated with the idea of Man as an integral part of Nature. The ecological movement generally uses the concept of sustainability in this way. The relationship between sustainability and our perception of reality has been examined by Alrøe and Kristensen who argue that the two meanings of sustainability arise from different attitudes to Nature. The resource sufficiency interpretation approaches Nature from the outside, whereas the functional integrity interpretation approaches Nature, in some sense, from the inside.The two different underlying perceptions of nature in resource sufficiency and functional integrity are Emphasized in Figure.

As an example of an farming rule that relates to sustainability as functinal integrety, we can mention the prohibited use of artificial fertilizers. By avoiding these fertilizers farming is forced to work with Nature's own nutrient cycles and become more self-reliant.

ANIMAL HUSBANDRY

In the EU regulations for organic animal production several general principles are given for the management and rearing of farm animals. In connection with an analysis of the ecological regulations relating to animal health and welfare much consideration was given to the question of whether farming involves distinctive animal welfare issues. The cease was that farming incorporates several distinctive aspects of animal welfare. First and foremost it relates to the ethical decision of how well animals should be kept, and associated with this, the definition of good animal welfare.

In farming animal welfare is seen to include better options for developing natural behaviour, these involving access to out-door areas and freedom of choice as a means of satisfying the preferences of individual animals. These options imply that care, as a concept of Man's responsibilities to animals, acquires greater importance. The detailed regulation of the layout of housing etc. does not necessarily ensure better animal welfare, since care and husbandry methods also play important roles. The "naturalness" of animals or lack thereof in farming can also be an important question, seen from the viewpoint of their integrity. In this context the breeding and reproduction of animals that are adapted to organic production systems, as well as the choice of breeding methods, are central concerns.

Finally farming incorporates the perception that Man and farm animals are parts of a wider ecological system. This perception can be of significance for animal welfare in the context of harmony, this concept expressing the sustainability, health and integrity of the system in the broadest sense. However, without doubt there are different values and perceptions associated with the concept of animal welfare, even within the ecological movement. Further clarification of this situation would provide a useful tool for the advancement of farming. In summary, one can imagine a forward-looking perspective for the development of farming in relation to animal welfare. Instead of demanding more detailed regulation, this perspective will focus on the special features of farming: greater opportunities for the development of natural animal behaviour and harmony in the structure and function of the farming system.

SOCIAL ASPECTS

Social issues take a central position in farming.

The following main objectives are set out in International Federation of Organic Agriculture Movements Basic Standards:

- To draw together the social and ecological consequences of organic production and processing.
- To ensure that all who work in farming and processing have a quality of life that meets their basic needs and provides adequate and satisfactory returns for their labours, including a good working environment
- To strive to make the whole chain of events in production, processing and distribution both socially and ecologically justifiable.

Table. Prioritised reasons for changing to farming

			Year of transition:			Reason for
-02	03-04	05-06	-2007	Avarage		transition:
Environmental	48	44	34	41	42	concerns
Disagree with conventional ageiculture	39	31	22	7	27	
Future of griculture	25	31	31	24	27	danish a
Agronomic	30	24	41	28	33	Challenge
Better foods	24	18	10	7	16	
Animal welfare	12	22	10	16	13	
Higher income	10	13	34	54	26	
Worries about working conditions	13	22	8	13	12	

n=923

In 2009 the Nordic ecological and biodynamic associations adopted the following principles:

- To produce foods of high quality, in sufficient quantities and fairly distributed
- To make it possible for the farmer to get a reasonable income, a safe working environment and meaningful work
- To strive for close contacts between food producers and consumers
- To ensure the greatest possible re-cycling of nutrients through the integration of urban societies, agro-ecosystems and natural ecosystems

There is thus no doubt that social issues take a central position in the ecological movement. On the other hand it must also be stated that these aspects are not particularly obvious in Danish farming at the present time. It is not included in Action Plan II – Ecology in Development. This can be taken to reflect the fact that in recent years disagreements have arisen in Denmark over the interpretation of these principles. At the ecological congress in Brædstrup there were also many contributions that focused on the social aspects of agriculture.

An essential reason for the dampening down of social issues in recent years is probably the rapid development that has occurred in the ecology movement. This has meant that many farmers have different reasons for going organic than had the original ecologists. Many of the new players – both within agriculture and associated activities – have but little contact with the currents and movements that brought about the today's ecological developments. This is Emphasized by the fact that in a recent study the reasons given by organic farmers for making the transition was seen to have changed in recent years. These farmers were asked to give their main reasons for transferring to farming. It can be seen that for many farmers environmental concerns were the main reason for their transfer to farming, both in the beginning and towards the end of the 1990s. But whereas many transitions in the early 1990s reflected a dislike of conventional farming methods, towards the end of the decade transfers were more often due to the potential for higher income.

The apparent falling disagreement with conventional agriculture is explained by the fact that the majority of new converts are organised within traditional farming organisations whereas the National Association for Farming currently represents relatively few of Denmark's organic farmers. The increased move to farming can thus be seen as an affiliation to a production concept rather than to a movement in which social aspects play a central role. Despite this development, however, there are still groups in Denmark that are orientated towards the local and original farming.

It can be difficult to interpret the fundamental principles that lie behind the social aspects of farming. Seen in relation to the other principles, especially those relating to solidarity with Nature and the concept of functional integrity, one can point to the concept of nearness as a central principle. We are here referring to vertical nearness, as exists between the players in the production, processing and distribution of organic foods, and horizontal or geographic nearness, in which the stream of nutrients is anchored locally in a naturally well-defined region. Michelsen and Kølster consider the nearness principle to be a very central priciple in farming.

SUMMARY OF FUNDAMENTAL PRINCIPLES

The main thread in the description of ecological principles and goals is that in farming Man is considered as an integral part of Nature, and that Nature is such a complex entity that we do not fully comprehend the consequences of our influence on it. From these fundamental assumptions some principles for action can be set out.

The cyclical principle is based on the fact that nutrients are recycled and used again, and with the help of sunshine renewable resources are built up. In the same way, Man must recycle nutrients, and avoid the use of non-renewable resources or ruthlessly exploit resources in general. Developments must be in harmony with Nature, and can be achieved by versatility and diversity in production.

The precautionary principle recommends care in the use of new technology, partly because Man is himself part of the natural cycle, and partly because Nature is complex and it can be difficult to foresee the consequences of Man's activities. A natural consequence of the precautionary principle is that older, known and well-functioning technologies will automatically have precedence over new technologies developed on a more theoretical basis. Similarly error-friendly technologies ought to be chosen rather than risky technologies. The nearness principle is concerned with how to secure special social aspects of farming transparency, safety, a sense of local belonging, and peace of mind, not to mention humanity and social justice. Direct contact between producer and consumer reduces the alienation that often characterises modern society. Learning on the basis of local experience and research into whole systems will be a central element for securing social and cultural values and the relationship to Nature.

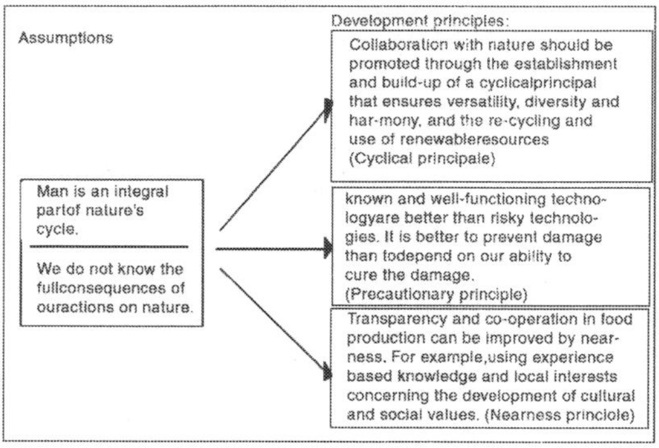

Fig. Fundamental Assumptions and Principal Relating to the Development of Farming.

The acquisition of information and knowledge about Nature and agriculture can be used to develop farming. Central to this is the importance of having a high degree of certainty about the possible consequences of using new technology, and an assurance that the development is in accord with cultural and social values.

TECHNOLOGICAL QUESTIONS RELATING TO THE DEVELOPMENT OF FARMING

There have been several statements about new tillage processes and the housing and treatment of farm animals.

A working hypothesis is that the development of farming depends very much on the acceptance or rejection of new technologies. Ideally, this choice is determined on the basis of principles and values. Any disagreements in principle can be revealed by examining accepted new technology and relating this to basic organic principles.

An attempt has been made to assimilate the last 10 – 15 years' debate on the use of new and older technologies that appeared in Farming.

Furthermore, an assessment is made of the extent to which this technique can be considered to be acceptable in relation to organic production.

PLANT PRODUCTION

Emphasizes that in the area of plant production there are many examples of the use of new technology in recent years. There are, however, certain exceptions, in particular the ban on Genetically Modified Organisms and synthetic sprays.

There has been, and there still continues, a debate on the steaming of top-soil and greenhouse production in compost. The steaming of topsoil does occur, however, in greenhouses.It is also evident that the debate has also extended to older technologies, such as spraying with sulphur and copper sulphate.

Table. Examples of techniques and inputs that are used and / or can be used inorganic plant production.

Technology: technology: pted	**New technology: debate**	**Acce- Banned**	**Under**	**Older debate**	**Banned**	**Accepted**	**Under**
Manually controlled row cultivation	X						
Mechanically controlled row cultivation		X					
Weeding robots	X						
Flame treatment	X						
Steaming of		X	X				top-soil
Straw covering	X	X					
Net covering	X						
Plastic covering	X						
Spraying with					X		sulphur
Spraying with sulphate						X	copper
Spraying with silicon preps.				X			humus and
Spraying with extracts				X			plant
Spraying with			X				chemicals
Kemink system cultivation)	X	X					(deep
Driving with machinery	X	X					heavy
Use of stubble /	X						catch crops
Use of compost				X	X		
Use of farmyard manure / liquid manure				X			
Use of slurry	X	X					
Use of artificial					X	fertilizers	
Greenhouse production without soil		X	X				
Traditional breeding				X			plant
Use of GMOs		X	X				
Use of micro	X	X					organisms

(EM)					
Use of c varieties			X	X	onventional
Use of manure			X	X	conventional
Use of fossil			X	X	fuel
Use of bio-gas	X	X			

ANIMAL PRODUCTION

In animal production there are only a few examples of new techniques that have been accepted without question. At the same time there are many older techniques that are under debate or banned. The limits of tolerance of new technology are evidently lower when it concerns individual farm animals than theyare for plants or the herd / flock as a whole.

Table. Examples of the technology debate in organic animal production

Technology: pted	New technology: Acce- debate	Under	Banned	Older technology: Banned	Accepted debate/		Under
Orientated to the individual:							
Dehorning of				X	X	cattle	
Castration of				X	X	cattle	
Ringing the				X		noses of bulls	
Castration				X	X	of pigs	
Ringing the				X	X	snouts of pigs	
Filing or teeth				X	X	clipping of	
Tail docking			X				
Beak trimming			X				
Curative – synthetics				X	X	medicines	
Curative non synthetics				X		medicines –	
Orientated to herds / flocks:							
Prophylactic medicines with synthetics			X				
Vaccinations				X	X		
Mechanical				X		milking	
Milking robots	X	X					
Use of breeds	X	X				traditional	
Use of insemination				X		artificial	
Use of technology	X	X	X			embryo	

Use of GMO			X	
Large herds	X	X	X	/ flocks
Covered		X	X	outdoor areas
No-straw			X	systems
No feeding of roughage/ zero grazing		X	X	
Feeding of			X	silage
Fully feeding	X			automated
Use of feeds and straw		X	X	conventional
Sale of bull	X	X		calves
Feeding with	X	X		green pellets
Loose housing X				

PROCESSING, MARKETING AND RE-CYCLING

Within the areas of processing, marketing and recycling, new technology is generally accepted for marketing, whereas the questions of processing and the recycling of town waste are subject to much greater debate.

Table. Examples of questions that are discussed in the areas of processing, distribution and sales,and the re-use of town waste

	New technology:				**Older technology:**		**Technology:**
Acce-	**Under**	**Banned**	**Accep-**		**Underpted**		**debateted**
debate/							**Banned**
Homogeni-		X	X				sation of milk
Pasteurisation				X	X		of milk / cheese
Additives		X	X				
Packaging	X	X					
Washing of vegetables and potatoes	X	X					
Farm gate and				X			market sales
Supermarket	X						sales
Sales by subscription arrangements	X						
Sales via the internet	X						
Long distance	X	X					transport
Re-use of human urine and compost				X			
Re-use of			X				sludge
Re-use of						X	human faeces
Work environment	X	X					

DISCUSSION ON THE USE OF NEW TECHNOLOGY

The ban on artificial fertilizers itself causes increased use of resources for the transport and handling of organic manures, as well as a dependence on conventional agriculture. It can be seen that in all the examples there is apparently broad acceptance of machine technology. Against this there is concern about or a ban on technology connected with synthetic inputs. The rationale behind this agrees well with the precautionary principle, since the use of machine technology is more limited and predictable in relation to Nature than are chemical and biological inputs.

The ban on artificial fertilizers cannot directly reflect the precautionary principle since the technique is well known, and its consequences predictable in relation to Nature. It can, however, be motivated by the cyclical principle, at least under Danish conditions in which large excesses of nutrients are sometimes used. The problem is, however, that in line with increasing specialisation in farming, the ban on artificial fertilizers itself causes increased use of resources for the transport and handling of organic manures, as well as a dependence on conventional agriculture. This can produce conflicts with other environmental issues, as well as with the nearness principle.

The use of organic manures, preferably in the form of well-rotted compost, is and has been considered to be most appropriate for farming. To a large extent this may relate to biodynamic farming principles in which the incorporation of oxygen and humus preparations into animal manures is seen as an important step for preserving and producing "living soil". Correspondingly, biodynamic farmers use other preparations, such as silicon, to promote the ripening process in corn. It must be recognised, however, that these farming practices are currently used to only a very limited extent in Denmark.

The use of slurry in farming has been the subject of much debate. Nevertheless, slurry-based systems now constitute a widely used and well-understood technology amongst organic farmers. The shift from compost to slurry rests with the accumulation of information over the last 15 years, several investigations having shown that animal manures, in the form of compost, have greater risk of loosing nitrogen than does slurry. Furthermore the fertilizer value of compost is much lower than that of slurry, particularly for the production of corn on sandy soils. It has not yet been shown that compost gives rise to better quality products and / or higher fertility.

A corresponding shift in technology is seen in cattle production, where silage is now widely used in farming. Silage is green material that is compacted and conserved under anaerobic conditions, in contrast to hay that is dried and stored aerobically.

The shifts in technology for the conservation and use of animal manures and feed are very fundamental, at least in relation to the biodynamic farming principles that formed an important backdrop to Danish farming. They provide a good example of the learning process that is taking place in farming. At the same time it is important to emphasise that this process is locally and nationally based on research findings and practical experience alike, the question of slurry use, in particular, having been widely discussed. From the middle of the '80s, farming in Denmark has been greatly inspired and influenced by organic production in southern Germany, Switzerland and Austria, where compost and hay-making were and are common practices. Gradually, as the organic movement in Denmark became more established, it adapted its technologies to Danish conditions.

In the area of animal husbandry, a number of older practices relating to individual animals have been debated in recent years. These practices can rightly be said to constitute an interferance in the nature and integrity of the animal, but they cannot be justified on the basis of the precautionary principle since they are well known, predictable and without risk to Nature. The background to this debate is thus the widespread Danish interest in animal welfare. In Denmark the salient attitude is regard for the animal according to a consciousness-centred ethic in which all conscious beings have a moral claim on respect. This attitude is not necessarily shared in many southern European countries.

At the present time the acceptance of these older practices for interfering in the nature and integrity of animals is unclear. On the other hand the fact that tail-docking and beak-trimming are forbidden is not an issue. This can be explained by the fact that the latter practices were primarily designed to counter the consequences of poor production systems, whereas other interventions are undertaken in respect to such factors such as the environment, product quality, and the safety of man and animal. Thus, at the individual level within farming the balance between respect for animals and concern for other matters is continuously being assessed. The special feature of farming is that, based on the precautionary principle, known and well-functioning practices are given precedence, whereas interventions relating to the size and efficiency of production are not accepted. In addition, consideration of other productions systems may perhaps instigate a new assessment.

With regard to developments at the herd level, several new technologies have been debated in recent years. Bans have been imposed on the use of GMOs and prophylactic medicines in accordance with the precautionary principle and the findings of risk assessments. Prohibitions have also been placed on production systems in which straw is not used for bedding, as well as systems that do not use roughages or allow access to out-door areas. These exclusions can be explained on the basis of the cyclical principle, since they ensure a relationship between forages and herds and thus the demand for harmony on the farm. Loose housing is a new method that is widely accepted in farming because it satisfies the natural behaviour needs of animals better than by the tie-stall system that constituted the most common type of housing at the beginning of the 1990s.In contrast to the situation in other countries, there is currently a great deal of learning associated with the introduction of new methods of processing and marketing to Denmark. This is explained by the different cultural backgrounds and the fact that the relationship of these approaches to Nature is much less intrusive than with plant production. It is difficult to imagine a ban on the washing of green vegetables and potatoes, since the technology is well known and its consequences for Nature predictable. In contrast to the situation in Sweden, there is a ban on the homogenisation of organically produced milk in Denmark, where this technique has been discussed and found to be unnecessary. In German-speaking countries there is much scepticism about Supermarket sales.

DISCUSSION AND PERSPECTIVES FOR THE DEVELOPMENT OF FARMING

FROM LOCAL SALE TO LARGE-SCALE ORGANIC OPERATION

That there is reasonably good accord between practical developments in farming and the precautionary and, to some extent, the cyclical principles, respectively. It can also be seen that a great deal of learning has occurred, and that many new techniques have been introduced which do not contravene its basic philosophy in the areas of weed control, the anaerobic storage of manure and feed, and the use of loose housing in place of the traditional tie-stall system.

At the same time it can be seen that virtually all the techniques accepted in Tables and have helped to raise the productivity of farming. These techniques have contributed to the increase of farm size and specialisation. The size of organic farms in Denmark has doubled from an average of 20 hectares in 1990 to 43 hectares in 2000; and these units are three times as big as the average of those in Switzerland and Austria.

As a consequence of larger farm sizes and specialisation particularly with respect to the cyclical and nearness principles. These developments will particularly affect physical distances and the concepts of versatility, variety and harmony in relation to Nature. In the Danish organic debate of the last 10 – 15 years, most emphasis has been placed on other parts, especially the material aspects of the cyclical and precautionary principles.

There are good reasons to believe that structural pressures will continue, particularly those associated with the increase in farm size and specialisation. Competition between organic farmers will increase this pressure, especially in a situation of over-production. Organic farmers should therefore consider the extent to which developments can be regulated or controlled with a view to maintaining confidence in this form of agriculture. In this context they should consider whether the nearness principle should be followed more explicitly.Using local and experience-base interpretations of the nearness principle, in earlier times organic farmers were more easily able to maintain an inner consistency and relationship in the development of production at the local level.

When products are disposed of locally customers become part of the local learning process. The problem is, however, that this basis for development does not harmonise well with the current rapid advance and internationalisation of organic agriculture. Free competition and the movement of products over long distances mean that quality symbols and the associated controls and regulations are the key to continued progress. To secure customer confidence and belief in organic products there must not be too many labels or quality symbols, and the arguments for choosing these products must be clear and unambiguous. There is thus a big need to re-interpret the nearness principle, such that its interpretation is expanded from the local to the national and international levels.

The alternative is to tighten up local conditions by banning sales via supermarkets, the internet, and other routes that do not involve direct trading and communication between the producer and the consumer. This alternative would reduce farming to a niche enterprise, however, since by far the greatest part of organically produced produce is sold without direct communication between these two groups. On the other hand, in the national and international perspective, Danish farming exploits the advantage of having only one organic trade mark that is well known and enjoys considerable confidence.

SIMPLIFIED BASIC CONTROL TO ENSURE PRODUCT CREDIBILITY

In the earlier development of farming, when there were fewer organic farmers and most were members of the same organisation, it was relatively easy to maintain a common organic philosophy based on the same values and principles. The modest size and identical affiliations of the organic group meant that there was not the same need to formulate common values and convert them to regulations since the physical proximity and informal social control between farmers was sufficient to secure reliable production. In association with the modern changes in farming, however, there is an increasing need to explicitly formulate these values such that they become more precise and better motivated. This is necessary because informal social control is no longer sufficient for ensuring credibility. Regulations must therefore be introduced as the basis of control, with the subsequent option of sanctions.In Denmark this process has involved extensive collaboration between the authorities and many interested organisations, and has resulted in the formulation of appropriate legislation and comprehensive regulations. The latter, and especially the controls, guarantee to the customer the credibility of organic production and its compliance with the rules of farming. In those situations where "inappropriate" conditions have been observed in organic production they have often been followed up with demands for further regulation and / or further control.

Amongst the most pressing motivations for a transition to farming in Denmark is the desire for a better environment, an agronomic challenge and more stable earnings. Criticism of conventional agriculture plays a much smaller role for recent converts than it did for the pioneers of the organic movement. Earlier, organic farmers were almost all members of the National Association for Farming, the membership of which also included many customers. Today the largest organisation for organic farmers is the Danish Farmers Union that primarily organises conventional farmers.Against this background there is no unequivocal picture of farming in Denmark. Amongst customers who are generally critical of conventional agriculture the distinction between conventional and farming can in some situations be blurred. This could have big implications for the credibility of organic produce.

There is thus a big need for clarity and simplification in our understanding of what farming really stands for, such that it comes to stand out as a real production alternative and not simply as a production method that complies with a set of more or less transparent regulations.

Relationships Between Producer and Customer

There is a need to further develop production, processing and marketing systems that are founded on close relationships between the producer and customer. These could include subscription arrangements and customer co-operatives in which the consumer is joint owner.

Openness and Transparency

There is a need for marketing systems that build on greater openness and transparency, in which the producer is no longer anonymous. For this, real information about the producer should be given on the packaging, or a particular route of sale should give information about the associated farms and invite customers to visit them.

Simpler and More Principled Regulations

Regulations of the type used in conventional farming or those that have little significance for farming should be excluded. Instead a few, simple regulations should be applied that emphasise the difference from conventional agriculture.

For example:

- The ban on artificial fertilizers, pesticides and GMOs

- The demand for grazing and roughages in animal production
- Increased withdrawal times for medicines
- Tighter controls on the importation of non-organic fertilizers and feed
- Rules that oppose specialisation
- Rules that promote the nearness principle.

The marketing systems described under a. and b. above can be considered as a modernisation of the nearness principle in which physical or geographical proximity is replaced by emotional nearness. They can therefore effectively advertise and symbolise farming. It should, however, be recognised that since we are talking of emotional nearness there is no guarantee that the remaining organic principles are respected, particularly those discussed under the cyclical principle, such as energy utilisation, versatility and the re-cycling of nutrients. Regulation is therefore the key for maintaining the credibility of organic produce, and the big challenge to research is to ensure that these regulations are founded on the best possible information.

2

Farming and Climate Change

AN OVERVIEW

Climate change will dramatically alter global food production. The inequity in food supply between industrialized and developing countries is expected to increase, as the 40 poorest countries in the tropical and subtropical zones will suffer most, both from droughts and periodic floods.

Agriculture is not only affected by climate change but also contributes to it. Ten to twelve percent of global greenhouse gas emissions are due to human food production. In addition, intensive agriculture has led to deforestation, overgrazing and widespread use of practices that result in soil degradation. These changes in land use contribute considerably to global CO2 emissions. Sustainable agriculture and food supply systems are thus more urgently needed than ever before. They must boost the capacity of agricultural production to adapt to more unpredictable and extreme weather conditions such as droughts and floods, reduce greenhouse gas emissions in primary food production and halt or reverse carbon losses in soils.

Organic agriculture is claimed to be the most sustainable approach in food production. It emphasizes recycling techniques and low external input and high output strategies. It is based on enhancing soil fertility and diversity at all levels and makes soils less susceptible to erosion. In this publication, farming and food systems are evaluated in the context of climate change scenarios. As simple answers cannot be given to such a complex and global problem, it is equally important to highlight recommendations for future development and research requirements in organic agriculture. Farming links productivity with ecology and creates livelihoods in rural areas: it is a surprisingly multifaceted option.

AGRICULTURE AS CAUSE AND VICTIM OF CLIMATE CHANGE

The current change in global climate is a phenomenon that is largely due to the burning of fossil energy and to the mineralization of organic matter as a result of land use. These processes have been caused by mankind's exploitation of fossil resources, clearing of natural vegetation and use of these soils for arable cropping. These activities have primarily led to a measurable increase in the carbon dioxide content of the atmosphere, an increase which results in global warming, as CO2 hinders the reflection of sunlight back into space, and thus more of it is trapped in the Earth's atmosphere.

Molecules of methane and nitrous oxide have a similar, but far greater effect: the global warming potential of methane is twenty times that of CO2, while that of nitrous dioxide is as much as 300 times greater. IPCC has published greenhouse gas emissions classified by different sectors. When calculating the climate impact of a certain production type it is always a question, where to put the cut-off points of a particular system. For instance which are counted under 'industry'. This needs to be taken into account when comparing farming systems. When considering the total food chain from the farm to the consumer, emissions from all the other sectors need to be included. Thus, the greenhouse gas emissions from all sectors related to agriculture may potentially sum up to 25-30% of all GHG emissions.

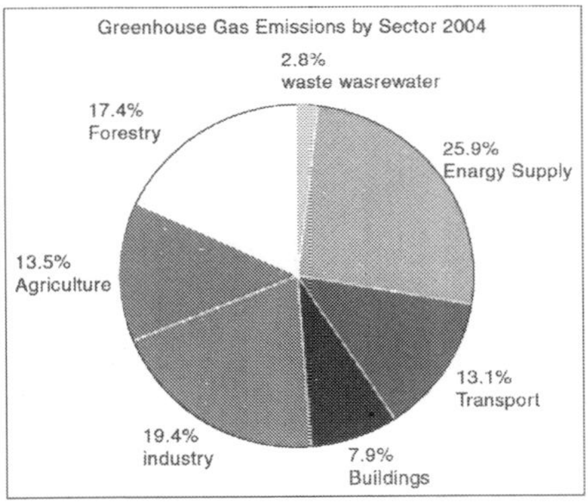

Fig. Greenhouse Gas Emissions.

GREENHOUSE GASES EMITTED BY THE AGRICULTURAL SECTOR

The Intergovernmental Panel on Climate Change, the annual amount of greenhouse gases emitted by the agricultural sector is estimated at between 5.1 and 6.1 gigatonnes CO2 equivalents2 in 2005. This represents approximately 10–12% of total greenhouse gas emissions.

Emissions of nitrous oxide originate mainly from:

- High soluble nitrogen levels in the soil from synthetic and organic nitrogen sources.
- Animal housing and manure management.

The main sources of methane emissions are:

- Enteric fermentation3 by ruminants.
- Anaerobic turnover in rice paddies.

- Manure handling.
- Compaction of soils resulting from the use of heavy machinery.
- Biomass burning, *e.g.* from slash-and-burn agriculture, emits both methane and nitrous oxide.

Vegetation – together with the soil ecosystem as the place for decomposition – generates large fluxes of carbon dioxide both to and from the atmosphere4. The Intergovernmental Panel on Climate Change, this flux is nearly balanced in agriculture with a net emission of 0.04 Gt CO2 equivalents per year, and represents less than 1% of global anthropogenic CO2 emissions.

Substantial emissions of CO2 from soils, however, originate from land-use changes such as deforestation. On the other hand, reforestation and afforestation are considered sinks for CO2.

By sequestering carbon dioxide in the soil, agriculture may contribute to the carbon cycle in a positive way. Whether the balance is positive or negative depends to a large extent on the cutoff points for the comparison as well as the management and farming practices applied. Agriculture has the potential to be a considerable CO2 sink, if good farming practices, like farming, are employed.

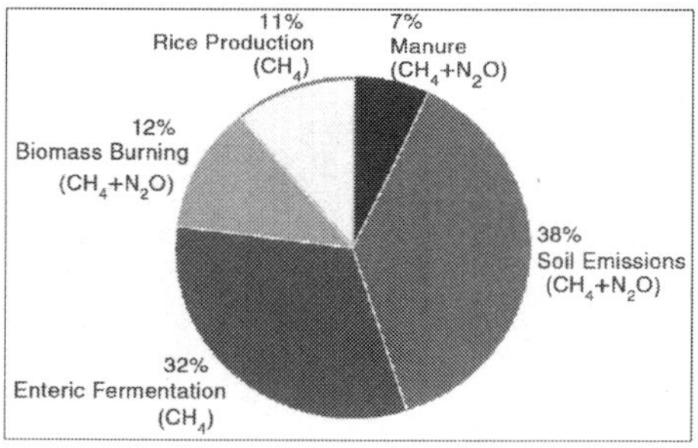

Fig. Greenhouse Gas Emissions in the Agricultural Sector in 2005.

FURTHER INCREASE IN EMISSIONS EXPECTED UNLESS AGRICULTURAL PRACTICES CHANGE

Predictions concerning the future global trends for greenhouse gas emissions from agriculture largely depend on physical and economic parameters that have a strong influence on total emissions. These parameters include: cost of fuel, economic development, evolution of livestock numbers, increase in productivity, new technology, availability of water, deforestation, and consumer attitudes and diet. Total greenhouse gas emissions from agriculture are expected to reach 8.3 Gt CO2 equivalents per year in 2030, compared to the current level of approximately 6 Gt CO2 equivalents annually.

HIGHER RISKS DUE TO MORE UNPREDICTABLE WEATHER

Current scientific models predict substantial environmental changes caused by increased emission of greenhouse gases. These changes will affect agriculture both in positive and negative ways. The forecast increase in global temperature of between 1.4°C and 5.8°C will result in alterations in precipitation patterns. Extreme weather events are expected to occur more frequently. Seasonal variations in weather events may pose risks to traditional methods of crop production either due to water constraints or surplus of water and erosion. Soil stability will become crucial in order to store water in the soil profile, to resist severe weather events and minimize soil losses. Vulnerable regions such as tropical and subtropical areas and high mountain regions are expected to suffer most from climate change.

MEASURES PROPOSED BY IPCC TO MITIGATE THE GLOBAL WARMING IMPACT OF AGRICULTURE

The Intergovernmental Panel on Climate Change IPPC has suggested a range of measures for mitigating greenhouse gas emissions from agricultural ecosystems. Ac cording to Barker et al., sink enhancement will contribute most to mitigation in this context. Prominent mitigation options in agriculture just as to IPCC include improved cropland management, improved grazing land management and the restoration of degraded soils.

THE POTENTIAL OF FARMING TO MITIGATE CLIMATE CHANGE

Agriculture can help to mitigate climate change by:

- Reducing emissions of greenhouse gases; and
- By sequestering CO2 from the atmosphere in the soil.

The potential of organic agriculture for both effects is high, as data gained from modelling both long-term field trials and pilot farms show.

REDUCTION OF GREENHOUSE GAS EMISSIONS

Farming has Lower Global Warming Potential

The global warming potential of agricultural activities can be defined as greenhouse gas emissions in CO2 equivalents per unit and area or per unit product. The global warming potential of farming systems is considerably smaller than that of conventional or integrated systems when calculated per land area. This difference declines, however, when calculated per product unit, as conventional yields are higher than organic yields in temperate climates. Under dry conditions or water constraints, organic agriculture may outperform conventional agriculture, both per crop area and per harvested crop unit.

Figure shows the results of comparative studies on the global warming potential of organic and conventional production. Organic farms do not only produce cash crops, but they use arable fields for temporary grass-clover and fodder production for livestock. A part of the grass-clover yields and nutrients is used to fertilize arable crops and soils.

Depending on the methodology and the calculations used, the results can differ substantially:

- In the two long term comparison experiments with arable rotations in Switzerland, the DOK trial and the Burgrain experiment the global warming potential of all crops was reduced by 18% in the organic plots. A reduction of warming potential has also been found in Dutch dairy farms and in some vegetable crops.
- In contrast, no difference or higher emissions for organic crops were modeled for the experimental farm in Scheyern, for 18 Bavarian commercial crop farms and for most of the vegetable crops from model farms in the Netherlands

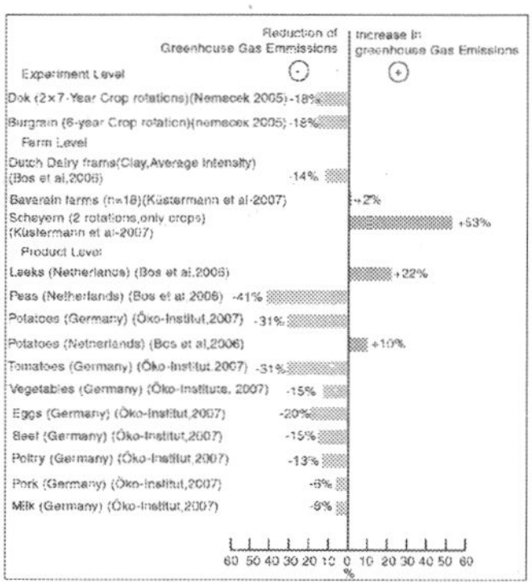

Fig. Relative Global Warming Potential of Organic Agriculture Compared to Conventional Agriculture. At the Experiment and Farm level, all Crops in a Rotation were Compared.

When losses and gains of soil carbon stocks are considered in the calculations, the global warming potential is considerably reduced for organic agriculture as shown in recent studies:

- *Scheyern experimental farm*: Decrease of 80%.
- *Bavarian survey of 18 commercial farms*: 26%.
- *Station experiment in Michigan*: 64%.

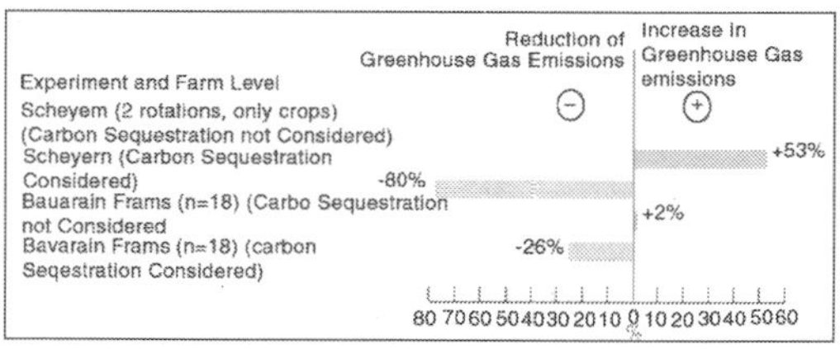

Fig. Relative Global Warming Potential of Organic Agriculture Relative to Conventional Agriculture with and without Consideration of CO2 Sequestration.

The Role of Nitrogen and N2O

Synthetic Nitrogen Fertilizers as a Major Contributor to Global Warming

The global warming potential of conventional agriculture is strongly affected by the use of synthetic nitrogen fertilizers and by high nitrogen concentrations in soils. Global nitrogen fertilizer consumption in 2005 was 90.86 million tonnes. It takes approximately 90 million tonnes of fossil fuel to produce this nitrogen fertilizer5. This is about 1% of global fossil energy consumption. In the UK, a 100-hectare stockless arable farm consumes on average 17,000 litres of fossil fuel annually through fertilizer inputs.

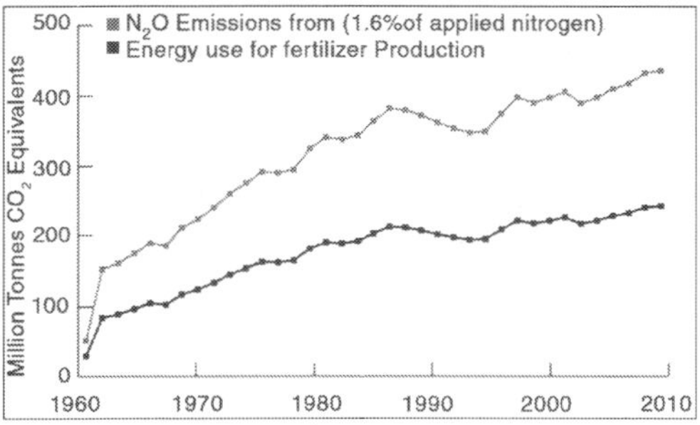

Fig. Energy used for Production of Synthetic Fertilizers and Emissions of Nitrous Oxide (N2O) from Soils6 after Application of Fertilizer, based on data from the International Fertilizer Industry Association IFA.

Organic Agriculture: Self-sufficient in Nitrogen

Organic agriculture, in contrast, is self-sufficient in nitrogen. Mixed organic farms practice highly efficient recycling of manures from livestock and of crop residues by composting. Leguminous crops deliver additional nitrogen in sufficient quantities. Badgley et al. calculated the potential nitrogen production by leguminous plants via intercropping and off-season cropping to be 154 million tonnes, a potential which exceeds the nitrogen production from fossil fuel by far and which is not fully exploited by conventional farming techniques.

Fig. Legumes.

Organic agriculture: Reduced Emissions of Nitrous Oxide

Emissions of nitrous oxide are directly linked to the concentration of easily available mineral nitrogen in soils. High emission rates are detected directly after fertilization and are highly variable. Denitrification7 is additionally enhanced in compacted soils. IPCC, 1.6% of nitrogen fertilizer applied is emitted as nitrous oxide. In organic agriculture, the ban of mineral nitrogen and the reduced livestock units per hectare considerably reduce the concentration of easily available mineral nitrogen in soils and thus N_2O emissions.

Furthermore, these factors add to lower emissions of nitrous oxide:

- Diversified crop rotations with green manure improve soil structure and diminish emissions of nitrous oxide.
- Soils managed organically are more aerated and have significantly lower mobile nitrogen concentrations. Both factors reduce emissions of nitrous oxides.

In the study by Petersen et al, lower emission rates for organic compared to conventional farming were found for five European countries. In a long-term study in southern Germany, Flessa et al. also found reduced nitrous oxide emission rates in the organic farm, although yield-related emissions were not reduced.

Integration of Livestock and Crop Production: An Important Contribution to Mitigation

The on-farm use of farmyard manure– a practice increasingly abandoned in conventional production – needs to be reconsidered in the light of climate change. While conventional stockless arable farms use synthetic nitrogen fertilizers, manure and slurry from dairy, beef, or from non-ruminant farms have become an environmental problem. In these livestock operations, nutrients are available in excess and overfertilization occurs. Emissions of CO_2, nitrous oxide and methane are likely to be very high and water pollution may occur when manure is treated as waste and not as recycled as a valuable fertilizer in the crops. Integration of livestock and arable production, the rule on organic farms, can thus reduce the global warming potential of food production. This fact is not calculated correctly in most global warming potential models, however, as livestock production is generally considered separately from crop systems.

Nitrogen Efficiency as a Key factor for the Reduction of Greenhouse Gases

Greenhouse gas emissions at farm level may be related either to the farm's nitrogen surplus or to the farm's nitrogen efficiency, as demonstrated by a scientific model of greenhouse gas emissions from European conventional and organic dairy farms. Farm nitrogen surplus can therefore be a good proxy for greenhouse gas emissions per unit of land. Since organic crop systems are limited by the availability of nitrogen, they aim to balance their nitrogen inputs and outputs and their nitrogen efficiency. Their greenhouse gas emissions are thus lower than those of conventional farming systems.

Methane Emissions

Methane accounts for about 14% of the greenhouse gas emissions. Twothirds of this are of anthropogenic origin and mainly from agriculture. Methane emissions stem to a large extent from enteric fermentation and manure management and in consequence are directly proportional to livestock numbers. Avoidance of methane emissions of anthropogenic origin and especially of agricultural origin is of particular importance for mitigation. Organic agriculture has an important, though not always superior, impact on reduction as livestock numbers are limited in organic farms.

The data available on methane emissions from livestock is limited, especially with respect to the reduction of GHG emissions from ruminants and manure heaps. Some authors suggest high energy feedstuff to reduce methane emissions from ruminants but the ruminants' unique ability to digest roughage from pastures would then not be used. Furthermore, meat and milk would be produced with arable crops where mineral nitrogen is an important CO2 emitter, and competition to human nutrition might become a problem.

Longevity of Animals on Organic Farms Contributes to Reduction of Methane Emissions

Organic cattle husbandry contributes positively to reducing methane emissions by aiming towards animal longevity.

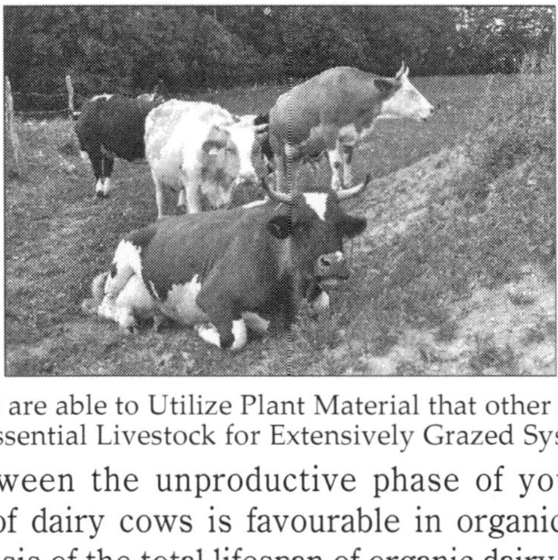

Fig. Since Ruminants are able to Utilize Plant Material that other Animals cannot, they are Essential Livestock for Extensively Grazed Systems.

The ratio between the unproductive phase of young cattle and the productive phase of dairy cows is favourable in organic systems because, calculated on the basis of the total lifespan of organic dairy cows, less methane is emitted. On the other hand, lower milk yields of organic cows caused by a higher proportion of roughage in the diet, might increase methane emissions per yield unit. A model calculation of the best yield-methane emission rate at different diets is missing. The slightly reduced yields of organic farms might be nearer the optimum than conventional dairy production.

Composting and Biogas Production as Measures for Mitigating Dlimate Change

Composting and biogas production are often suggested as measures for mitigating climate change. In the context of climate change, the benefits of aerobic fermentation of manure by means of composting are ambiguous, as a shift from anaerobic to aerobic storage of manure can reduce methane emissions, but will increase emissions of nitrous oxide by a factor of 10. A very promising option, however, is controlled anaerobic digestion of manure and waste combined with biogas production. While this option is not restricted to organic production methods, organic agriculture has been at the forefront of biogas production systems for decades. Attention must be paid however to the economic viability of biogas production systems.

Methane Emissions from Organic rice Production and from Ruminants: Improved Techniques needed

Methane emissions from organic rice production and ruminant production do not differ substantially from those of conventional production. Better rice production practices in organic and conventional agriculture, such as avoiding continuous flooding or choosing low methaneemitting varieties could enhance reduction of methane emissions. The multi-target approach of organic farmers and the fact that they are often more highly skilled could enhance implementation of improved production techniques.

FARMING SEQUESTERS CO2 IN THE SOIL

Soil Erosion Results in Loss of Soil Carbon

Arable cropland and permanent pastures lose soil carbon through mineralization, erosion and overgrazing. Global arable land loss is estimated at 12 million hectares per year, which is 0.8% of the global cropland area. This rapid loss is confirmed by experimental data from Bellamy et al. (2005) in England and Wales. Between 1978 and 2003, they found carbon losses in 92% of 6000 soil samples. Annual CO2 emissions from intensively cropped soils were equivalent to 8% of national industrial CO2 emissions.

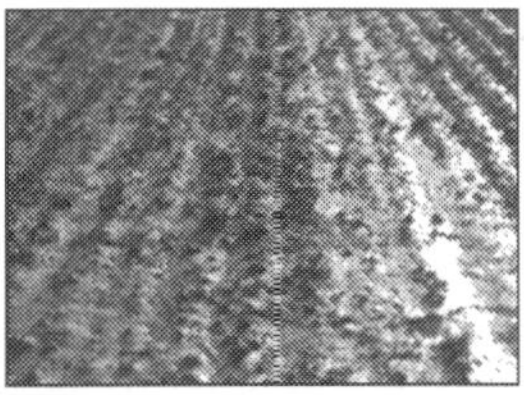

Fig. The Better Drainage and Water-holding Capacity of Organic Soils Reduces the risk of Drought and Soil Erosion.

If agricultural practices remain unchanged, the loss of organic carbon in typical arable soils will continue and eventually reach a new steady state at a low level. The application of improved agricultural techniques, however, stops soil erosion and converts carbon losses into gains. Consequently, considerable amounts of CO2 may be removed from the atmosphere.

Organic Land Management: Carbon Gains

Organic land management may help to stop soil erosion and convert carbon losses into gains particularly due to:

- The use of green and animal manure.
- Soil fertility-conserving crop rotations with intercropping and cover cropping.
- Composting techniques.

Higher Soil Organic Matter Content in Farming

Farm comparison and long-term field trials show that organically managed soils have significantly higher organic matter content. It is estimated that, under Northern European conditions, conversion from conventional to farming would result in an increase of soil organic matter of 100 to 400 kg per hectare annually during the first 50 years. After 100 years, a steady state, *i.e.* a stable level of soil organic matter, would be reached.

Carbon Sequestration rates on Organic Farms

Under 'real' conditions in long-term experiments, or in farm-level comparisons, carbon sequestration rates vary considerably.

Organic Versus Conventional Soil Conservation Strategies: No-tillage and Minimum-tillage Cropping

In the past decade, agricultural techniques have been developed to maintain soil fertility and soil quality.

By reducing the intensity of tillage, soil conservation can be improved, and water and wind erosion can be considerably reduced.

- *Robertson et al*: compared the greenhouse warming potentials of conventionally tilled, no till and farming systems in the Mid Western US and found none of these agricultural systems to be climate neutral. Whereas no-till reduced the GWP of conventional tillage by 88%, organic production with legume cover was only 64% lower than conventional tillage.

- In a nine-year system comparison experiment in Beltsville, it was shown that the farming approach provided excellent soil fertility building and was superior to conventional no-tillage techniques, despite the use of a plough.

No-tillage cropping is mainly practised on stockless farms, which leads to highly specialized farms – either crops or animals – and excess manure on the animal farms becomes an environmental problem. Nitrate excess in the soil triggers emissions of nitrous oxide, as well as nitrate leaching and phosphorus run-off. The organic approach involving local recycling and nutrient use in a mixed-farm approach offers many ecological benefits. In very fragile soils, it is Non-etheless recommended to use minimum-tillage techniques in farming as well. Several research projects in different parts of the world are working on such systems. For instance, in Switzerland a long-term trial was recently started that analyses the effect of reduced tillage on crop yields and weed infestation. Similar research projects are running at Bonn University in Germany. Technically, there is no inherent incompatibility between organic and minimum-tillage cropping.

Agroforestry, Permaculture, Polyculture

Agroforestry is a management system that integrates perennial and annual crops in a twocanopy or multi-canopy production system. This guarantees better exploitation of light, water and soil nutrients and protects soil more effectively from erosion and leaching. It leads to a more diversified and sustainable production system than many treeless alternatives and provides increased social, economic and environmental benefits for land users. In the humid tropics, agroforestry is seen as viable alternative to slash-and-burn agriculture.

Fig. Agroforestry System in Cuba.

The CO2 sequestration potential of agroforestry in the short and medium term is mainly above ground. The additional carbon in the standing vegetation may increase by 50 tonnes per hectare; in the soil by an additional 7 tonnes per hectare. These increases were measured 20 to 25 years after recultivation of cleared forests.. Even though agroforestry is neither restricted nor exclusive to organic agriculture, organic principles suit it very well. Consequently, organic agriculture could play a role in the development of agroforestry systems and combining these two systems is a potential solution for reducing greenhouse gases, sequestering carbon dioxide and increasing the productivity of agro-ecosystems.

MITIGATION POTENTIAL OF ORGANIC AGRICULTURE BEYOND PURELY AGRICULTURAL PRACTICES

Changing Consumer Behaviour and Diet

The greatest potential for reducing greenhouse gas emissions from agriculture would be to change consumer behaviour. Production of meat requires inputs that are seven times as high as the inputs needed to produce the same quantity of non-meat calories.

Organic agriculture aims at precisely this goal: consumption of less-processed products and increased consumption of products like cereals, potatoes, pulses and oils. Greenhouse gas emissions are highest in beef production followed by pork, poultry and egg production and milk. Emissions from production of plant foods are generally below 500 g CO2 equivalents per kg. Although ruminants are major methane emitters, they are crucial to global food security as they tap into an area of 3,432 million hectares worldwide that is not suitable for crop production.

These animals carry bacteria in their rumen that make plant material digestible, that other animals are hardly able to use, but unfortunately this fermentative process is also emitting methane.

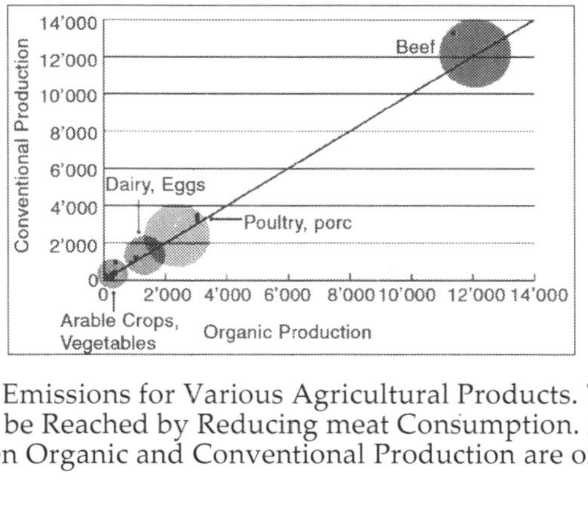

Fig. Greenhouse Gas Emissions for Various Agricultural Products. The Major Reduction of Emissions could be Reached by Reducing meat Consumption. In human diets the Differences Between Organic and Conventional Production are of Minor Relevance.

Stopping Deforestation

In farming, preparation of the land by burning vegetation is restricted to a minimum. Farming thus contributes to halting deforestation with its highly negative impact on climate change. Often, the opposite argument has been made, as organic agriculture usually needs more land to produce the same amount of food as by conventional farming. This might be compensated by the potential of organic agriculture for aiding reclamation and making use of degraded land due to its favorable effects on soil fertility and soil organic matter. In addition careful land use and management as in farming enhances environmental security and will help to stop losses of fertile arable land not only by erosion.

DOES FARMING HAVE GREATER POTENTIAL TO ADAPT TO CLIMATE CHANGE

Agricultural production in most parts of the world will face less predictable weather conditions than mankind experienced during the intensification of agriculture over the last century. Weather extremes will become predominant. Resilience and adaptiveness are new requirements gaining importance for innovation in agriculture. The Intergovernmental Panel on Climate Change defines adaptation to climate change as 'adjustment in natural or human systems in response to actual or expected climatic stimuli or their effects, which moderates harm or exploits beneficial opportunities'.9

TRADITIONAL SKILLS AND KNOWLEDGE AS A KEY TO ADAPTATION TO CLIMATE CHANGE

Traditional skills and knowledge have been neglected in intensive agriculture, although they are now being partially recaptured by integrated pest management. Organic agriculture, on the other hand, has always been based on practical farming skills, observation, personal experience and intuition. Knowledge and experience replaces or reduces reliance on inputs. This knowledge is important for manipulating complex agro-ecosystems, for breeding locally adjusted seeds and livestock, and for producing on-farm fertilizers and inexpensive nature-derived pesticides. Such knowledge has also been described as a 'reservoir of adaptations'.

ORGANICALLY MANAGED SOILS ARE BETTER ADAPTED TO WEATHER EXTRE

Farming practices such as organic agriculture that preserve soil fertility and maintain or even increase organic matter in soils are in a good position to maintain productivity in the event of drought, irregular rainfall events with floods, and rising temperatures.

Soils under organic management retain significantly more rainwater thanks to the 'sponge properties' of organic matter:

- These 'sponge properties' were described for heavy loamy soils in a temperate climate in Switzerland where soil structure stability was 20–40% higher in organically managed soils than in conventional soils
- The amount of water percolating through the top 36 cm was 15–20% greater in the organic systems of the Rodale farming systems trial compared to conventional systems. The organic soils held 816,000 litres per ha in the upper 15 cm of soil. This water res ervoir was likely the reason for higher yields of corn and soybean in dry years.
- It was found that water capture in organic plots was twice as high as in conventional plots during torrential rains. This significantly reduced the risk of floods, an effect that could be very important if organic agriculture were practised on much larger areas.

ENHANCING PRODUCTIVITY OF DEGRADED SOILS BY BUILDING SOIL FERTILITY

Experience with degraded soils of the arid tropics have shown that agricultural productivity can be enhanced using soil fertility building techniques. In the Tigray province of Ethiopia, one of the most degraded parts of the country, agricultural productivity was enhanced by soil fertility techniques such as compost application and introduction of leguminous plants into the crop sequence. By restoring soil fertility, yields were increased to a much greater extent both at farm and regional level than by using bought mineral fertilizers. This large-scale experiment underlines the importance of organic matter and soil fertility for ensuring productivity in dry regions and partly explains the surprisingly high yields from organic crops found by Badgley et al..

DIVERSITY ENHANCES FARM RESILIENCE

An additional strength of farming systems is their diversity – including the diversity of crops, fields, rotations, landscapes and farm activities. The high level of diversity of organic farms provides many ecological services that significantly enhance farm resilience. Positive effects of enhanced biodiversity on pest prevention have been shown by several authors. Similar effects of diversified agro-ecosystems on diseases and better utilization of soil nutrients and water are likely to occur.

WEAKNESSES OF ORGANIC AGRICULTURE IN THE CONTEXT OF CLIMATE CHANGE

FARMING IS LESS PRODUCTIVE

Agronomically difficult crops as a challenge

One major criticism of organic agriculture is that productivity is lower compared to intensive conventional agriculture. Under geoclimatic conditions that allow for a very high yield, in the case of some crops the relative advantage of organic agriculture in terms of energy consumption per land area compared to conventional production may switch to the contrary when calculated on the basis of crop or livestock yield. This is particularly true in the case of highly demanding crops such as potatoes, grapes fruits and horticultural crops especially from greenhouse production. Pest, disease and weed management problems relating to these crops have not yet been resolved satisfactorily. Consequently, the yield for these crops is too low and the energy input, even though relatively low on a land area basis, becomes relatively high when applied to a crop unit. With more research into organic agriculture, however, progress may be expected in this field.

Better Technology Transfer could Improve Organic Yields

The productivity of organic agriculture is often underestimated by many scientists and policymakers. Organic agriculture represents a very productive food supply system that relies on recycling strategies. Badgley et al. modelled the yields stated in 293 on-farm and onstation publications and concluded that, compared to high-input agriculture in developed countries, the average yields in organic crop and livestock production are 92% of those in conventional agriculture.

Table. Organic crop yields as a percentage of coventional reference yields

	CH	AT	DE	IT	FR
Wheat	64-75	62-67	58-63	78-98	44-55
Barley	65-84	58-70	62-68	55-94	70-80
Oats	73-94	56-75		88	
Grain Maize	85-88		70	55-93	66-80
Oilseeds	83	78-88	60-67	48-50	67-80
Potatoes	62-68	39-54	54-69	62-99	68-79

As scientific progress in conventional agriculture has accelerated in recent years, if we look only at the most recent data and at selected countries in temperate climate zones, yield differences as obtained from official databases are generally higher. In the case of wheat organic yields vary between 58% and 98% of those of the conventional pair. This rather wide range of yields measured by farm surveys indicates that yields on organic farms might be improved considerably, *e.g.* by better technology transfer from research to practice.

Farming Performs Better Under Water Constraints

Interestingly, yields from organic agriculture under conditions where water is limited during the growing period, and under subsistence farming, are equal or significantly higher than those from conventional agriculture. A comparison of 133 studies from developing countries concluded that organic plant and livestock yields were 80% higher than their conventional counterparts. In temperate climate zones higher yields for maize and soya were registered in organically managed fields in dry seasons. The available data indicates that the technique inherent to farming of investing in soil fertility by means of green manure, leguminous intercropping, composting and recycling of livestock manure could contribute considerably to global food productivity. Further improvements in manure storage and application techniques are also required, however, in organic systems in order to reduce nitrous oxide and methane emissions.

HIGH DEPENDENCY ON NUTRIENTS DERIVED FROM LIVESTOCK

Some critics are concerned about the dependency of organic cropping upon nutrients deriving from livestock. This criticism, however, underestimates manure as a valuable and potentially useful resource. Moreover, this is not a relevant weakness since the numbers of animals kept in agriculture depends mainly on consumer demand. In order to reduce greenhouse gas emissions, efficient and direct recycling of manure and slurry is the best option, since it avoids long-distance transport and consumption of energy for synthetic fertilizer production. The combination of crop and livestock production is currently the most efficient way of bringing organic 'waste' from livestock production back into the carbon stock of the soils and use it as a locally available resource for crop fertilization and enhancing soil quality.

When integrated into arable farming systems, ruminants exploit leguminous crops and intercrops that are needed to produce nitrogen, provide soil cover and capture soluble nutrients, building up soil fertility and soil structure. From the point of view of integrating ecology and sustainable resource use, the combination of organic cropping with livestock production is undoubtedly a strong point.

MORE FUNDING IS NEEDED FOR RESEARCH ON FARMING

As 99 per cent of the world's public and private research funds have focussed on optimizing conventional and integrated food and farming systems during the last decades, major progress and solutions can be expected as a result of agro-ecological and animal welfare research activities.

CLIMATE CREDIT FOR FARMING

Agriculture, both organic and conventional, has the potential to make a cost-effective contribution to mitigation. The Intergovernmental Panel on Climate Change estimates that agricultural greenhouse gas mitigation options are costcompetitive with non-agricultural options in achieving long-term climate objectives.

SPECIAL BENEFITS OF ORGANIC AGRICULTURE

Within agriculture, organic agriculture holds an especially favourable position, since it realizes mitigation and sequestration of CO2 in an efficient way. Compared to other agricultural systems, farming is a well-defined system that is already based on certification and that could easily be extended to meet the standards of CDM. Organic production has great mitigation and adaptation potential, particularly with regard to soil organic matter fixation, soil fertility and water-holding capacity, increasing yields in areas with medium to low-input agriculture and in agroforestry, and by enhancing farmers' adaptive capacity.

Paying farmers for carbon sequestration may be considered a win-win-win situation as:

- CO2 is removed from the atmosphere
- Higher organic matter levels in soil enhance their resilience; and
- Improved soil organic matter levels lead to better crop yield.

Improvement is needed, however, with regard to yields and methane emissions. Organic agriculture, with its holistic multi-target approach, offers further relevant advantages with regard to lifestyle changes primarily in developed countries. Its numerous cobenefits could greatly assist the development of rural societies in southern countries. It would benefit the cause if agriculture, and in particular organic agriculture, could be included as a high-benefit / low-cost CO2 reduction system in the next climate agreement, negotiations on which, referred to as the 'post 2012 negotiations', are about to begin.

CO2 SEQUESTRATION EXCLUDED FROM THE CLEAN DEVELOPMENT MECHANISM

Agriculture, however, and organic agriculture in particular, are barely able to benefit from climate credits, of which the Clean Development Mechanism is the most popular. Even though agriculture could apply for CDM credits for reduction of methane through biogas installations for example, the single most important measure in terms of greenhouse gas reduction, sequestration of CO2 in soils, is excluded from the CDM mechanism. Likewise, the 'gold standard' launched by the World Wide Fund for Nature formally excludes agricultural CO2 sequestration sinks from eligibility for carbon credits.

For both organizations, the reason is that agricultural CO2 sinks are considered to be temporary in nature only. Once CO2 has been sequestered, a change in land use poses the risk that a substantial amount of CO2 will be released back into the atmosphere. The logic of this perception leads to a situation where only afforestation or reforestation projects are supported. Organic mixed farms, especially when good compost is used, tend to build up organic matter with a higher proportion of stabile fractions. Therefore, quitting farming might not lead to a faster release of CO2 than cutting a tree plantation.

BENEFITS OF CO2 SEQUESTRATION ACKNOWLEDGED BY IPCC

The Intergovernmental Panel on Climate Change and other authors, however, state that CO2 sequestration in agricultural systems, agroforestry, improved land-use systems or reclamation of degraded soils for cultivation have many benefits.

The most important of these benefits are:

- The fact that sequestration could take effect very quickly – buying us time, as is noted often, since initial sequestration peaks can be reached after 5 to 10 years.
- CO2 sequestration in agriculture is very cost-effective and, as a result of its cobenefits, would also greatly assist rural development.

VOLUNTARY CO2 EMISSIONS MARKETS

Due to these incontestable benefits, voluntary CO2 emissions markets have emerged.

The best known of these are probably:

- The World Bank's BioCarbonFund
- The Chicago Climate Exchange
- The European Climate Exchange

Emission certificates are traded at prices that are currently lower than those suggested by IPCC so as to achieve a substantial contribution. Because the market is voluntary, it lacks the security of the 'official' Kyoto Protocol and is consequently somewhat speculative in nature. Strict and technically sound guidelines for the execution of voluntary carbon reduction projects will greatly assist the standardized implementation of voluntary carbon CO2 projects. This will increase projects' credibility, transparency as well as traceability and tradability. Standards like those of the Voluntary Carbon Standard published in November 2007 use the existing Clean Development Mechanism methodology where possible, but they broaden the scope by including additional methods as well as land-use systems to reduce or capture CO2.

CEASE

BENEFITS OF FARMING

The benefits of farming regarding climate change can be summarized as follows:

- Organic agriculture has considerable potential for reducing emissions of greenhouse gases.
- Organic agriculture in general requires less fossil fuel per hectare and kg of produce due to the avoidance of synthetic fertilizers. Organic agriculture aims at improving soil fertility and nitrogen supply by

using leguminous crops, crop residues and cover crops.

- The enhanced soil fertility leads to a stabilization of soil organic matter and in many cases to a sequestration of carbon dioxide into the soils.
- This in turn increases the soil's water retention capacity, thus contributing to better adaptation of organic agriculture under unpredictable climatic conditions with higher temperatures and uncertain precipitation levels. Organic production methods emphasizing soil carbon retention are most likely to withstand climatic challenges particularly in those countries most vulnerable to increased climate change. Soil erosion, an important source of CO2 losses, is effectively reduced by organic agriculture.
- Organic agriculture can contribute substantially to agro forestry production systems.
- Organic systems are highly adaptive to climate change due to the application of traditional skills and farmers' knowledge, soil fertility-building techniques and a high degree of diversity.

WEAKNESSES OF FARMING AND RESEARCH REQUIREMENTS

This paper recognizes that organic agriculture also has weaknesses, mainly related to productivity and yield losses in some crops and production areas. Such issues highlight the need for research. Total European research funding for organic agriculture currently represents less than 1% of the total food and agriculture research budget.

In order to improve organic agriculture's performance and to allow more assistance to be provided to organic agriculture projects in low-input or developing countries, where CO2 mitigation would be most beneficial:

- Soil fertility management, crop growth and health.
- Better exploitation of leguminous plants in improved crop sequences.
- Habitat management with improved manipulation and exploitation of diversity at all levels.
- Crop breeding programmes focusing on the adaptability of plants to low-input situations in soils, on weed competition, and on pest and disease tolerance.
- Improved plant protection techniques and compounds from natural sources.
- Breeding strategies and programmes for adaptability to management and environmental stress situations in organic livestock production.
- Reduced tillage organic systems.

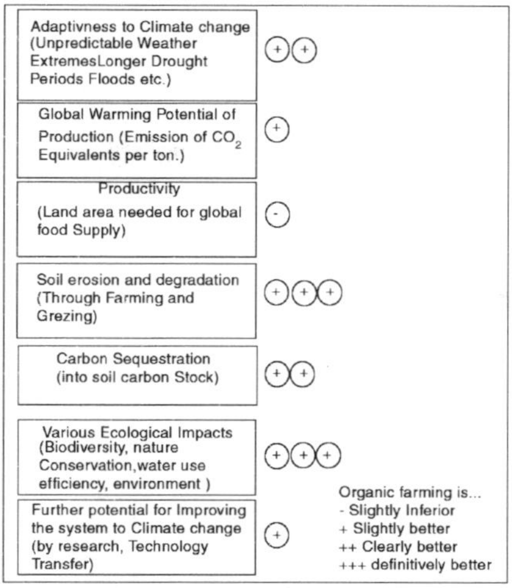

Fig. Performance of Organic Agriculture Compared to Conventional Agriculture in the Context of Climate Change.

In spite of these weaknesses, organic agriculture is so far the most promising approach for mitigation and adaptation to climate change. Organic agriculture represents a positive example of how farmers can help mitigate climate change and adapt to its predictable and unpredictable impacts. It can serve as a benchmark for allocating development resources to climate change adaptation, or to measure progress in implementing climate-related multilateral environmental agreements.

3

Gardening and Farming Techniques

AQUAPONICS

Aquaponics is the symbiotic cultivation of plants and aquatic animals in a recirculating environment. Aquatic animal effluent accumulates in water as a by-product of keeping them in a closed system or tank. The effluent-rich water becomes high in plant nutrients but this is correspondingly toxic to the aquatic animal.

Plants are grown in a way that enables them to utilize the nutrient-rich water. The plants take up the nutrients, reducing or eliminating the water's toxicity for the aquatic animal. The water, now clean, is returned to the aquatic animal environment and the cycle continues. Aquaponic systems do not discharge or exchange water. The systems rely on the relationship between the aquatic animals and the plants to maintain the environment. Water is only added to replace water loss from absorption by the plants, evaporation into the air, or the removal of biomass from the system.

Aquaponic systems vary in size from small indoor units to large commercial units. They can use fresh or salt water depending on the type of aquatic animal and vegetation.

HISTORY

Ancient Roots

Some consider Aztec chinampas to be the first example of aquaponics. Others refer to ancient Egypt. Either way, it is clear that aquaponics have ancient roots."Integrated vegetable growing and fish farming polyculture systems have long been used in Far Eastern countries such as China and Thailand. Farm wastes are commonly added as feed to fish ponds and fish are often cultured in flooded rice paddies."

Aquaponics in the United States

Many institutions and enterprises followed on the efforts at North Carolina

State University ; notably by the University of Arizona Environmental Research Labs, NASA/CELSS, S&S Aquafarms, The Freshwater Institute, University of Arkansas, Bioshelters, Inc, Global Aquatics, Inslee Fish Farms and others who carried out 'research and development' of aquaponics.

From the 1980s to present day the two distinct aquaponic systems are:

- "Deep Water" or "Raft Culture" aquaponics which is the primarily research carried out at the University of the Virgin Islands under the guidance of Dr. James Rakocy and;
- Reciprocating aquaponics based on the techniques developed by Mark McMurtry, et al. at NCSU

The University of the Virgin Islands Aquaculture Programme has developed an aquaponic system through over 20 years of research into its design and operation. The system can produce over 10,000 lbs. of tilapia annually and a variety of vegetables that are harvested weekly in staggered production or as needed by other fruiting crops The aquaculture Programme promotes several principles of aquaponics that can be applied to any size system, from hobby-scale to commercial-scale.

These principles include:

- System design that balances feed input to vegetable growing area,
- Constant input of feed by staggering fish stocking,
- Constant nutrient uptake by staggering vegetable production,
- Continuous water flow and
- Maintaining pH of 7.0. Other guidelines also apply and lead farmers to productive and profitable enterprises.

The University of the Virgin Islands teaches a course each June, the "International Aquaponics and Tilapia Aquaculture Course" to participants from around the globe. These individuals return home to develop their own commercial enterprises based on the aquaponic principles taught in the course.

In 1997 Rebecca Nelson and John Pade began publishing the Aquaponics Journal. The Aquaponics Journal was the first publicaiton to bring the research and various applications of aquaponics to people around the globe. Nelson and Pade also began offering training and internships at their facility in Mariposa, CA during the 1990s. In 2004 they produced the first video on aquaponics and in 2008, they wrote and published the first comprehensive book on aquaponics, "Aquaponic Food Production." They are known throughout the industry as leaders for offering training, consulting and sytstems. Nelson and Pade work closely with Dr. Rakocy of the University of the Virgin Islands to bring the research on aquaponics into mainstream agriculture. Now located in Montello, WI, they are expanding their business and are the only full-spectrum provider of aquaponic systems, supplies, training and consulting in the US. A graduate of the 2007 University of the Virgin Islands Short Course in Aquaponics, Tim Mann and his wife and partner Susanne started an aquaponics farm in Hawaii in 2007. This was the first aquaponics farm in the USA to have its produce USDA Organic certified and Food Safety Certified, and is currently shipping 1,600 pounds of its organic lettuce mix and 300 pounds of white tilapia per month. Friendly Aquaponics teaches the "Commercial Aquaponics Training" twice a year to participants worldwide. Attendees have developed their own commercial and educational enterprises based on the information provided in the course.

Aquaponics Research in Canada

The first aquaponics research in Canada was a small system added onto existing aquaculture research at a research station in Lethbridge. Later, a larger set-up was built in Brooks, Alberta. Scientists, especially Dr. Nick Savidov, at this research station have made some interesting findings related to rapid root growth in aquaponics systems.

Dr. Savidov's research is unusual for aquaponics as it is based out of a plant science background. His team took the system developed at the University of Virgin Islands and adapted it into a system in an Alberta greenhouse. One of the interesting adaptations that he made was to run his system at a low pH level, rather than a system that is half-way between the ideal pH for plants and neutral to high pH for fish. He found that even with a lower pH, the fish could survive nicely because of other advantages in the system over traditional aquaculture.

Commercial Aquaponics in Canada

Many of the pioneers of aquaponics in Canada are no longer in production. ounded in 1998, Future Aqua Farms in Nova Scotia is widely considered the first commercial aquaponics project in Canada. North Atlantic Aquaponics Ltd., in Newfoundland apparently grew eels in their aquaponic system., Agua Aquaponics Greenhouses International, in Manitoba is another early project that is no longer operational.

Other aquaponic systems tried aquaponics and then switched Towards pure aquaculture. MDM farms in Alberta switched from a small aquaponics system back to a pure aquaculture system. Greenview farms also in Alberta, switched from aquaponics to a pure aquaculture system at the same time that it changed its name to Aquagrow Enterprises International Ltd.

Only a few commercial aquaponics systems continue operations in Canada. Cultures Aquaponiques M.L. Inc, in Quebec is an early aquaponics system in Canada, and an unusual one in that it combines trout and floating lettuce production. Circle M Trout Farm, in Alberta also uses trout, but does not recirculate their water because they use it to produce fruiting vegetable crops. They run the trout water through the greenhouse in pipes to warm it up and then into the plant beds. After the plants have taken the nutrients out of the water, the water is too warm to recirculate and it is taken out of the system. Floating Gardens Ltd., in Saskatchewan won an award for their business plan using aquaponics and advertise a start-up date of summer 2009. They plan to grow tilapia and various vegetables for the Saskatchewan market.

Aquaponics in the Barbados

The easternmost Caribbean island nation of Barbados is considered to be the most developed of the Caribbean states, having one of the highest per capita incomes in the region.The densely populated island nation of Barbados has limited resources required for a prosperous agricultural industry. This high population density and seasonal influx of foreign tourists - over 570,000 tourists stayed in Barbados in 2007 and over 600,000 cruise ship passengers visited the country that same year – places stress on the country's key national resources. Barbados is known as the 15th most water scarce country in the world and freshwater withdrawal per capita is 333 m3/year.

The long history of intensive plantation-style monoculture production over the past 300 years in Barbados has made extensive contributions to soil quality problems in Barbados. These include the erosion of topsoil, a decrease in soil fertility, and the consequent application of large amounts of fertilizer and pesticides in order to maintain productivity. The colonial history of Barbados has left behind a reliance on monocrop, plantation-style agriculture which focuses on the production of a single cash crop in large amounts. The agricultural industry in Barbados therefore still consists primarily of sugar cane cultivation and the sugar, rum and molasses production industry.Production of food crops in the country is quite low as Barbados imports around 80% of its food, including large amounts of fruits and vegetables.

Only 10% of the labour force in Barbados is involved in agricultural activities as agriculture must compete with more profitable industries and forms of land use such as the growing tourism and real-estate sector. The reliance on outside factors and world markets associated with such high levels of food importation place the country in a position of dependency and hinders progress towards self-sufficiency in terms of food production.

Traditionally, small-scale farming faces challenges regarding the necessity to incorporate high-input technologies into their production in order to be able to compete on the global market. These technologies involve high costs and significant initial investments that cannot be afforded by all rural food producers and which increase production costs and therefore the cost of locally produced crops. This makes small farmers uncompetitive against cheaper imported items. Aquaponics has the potential to lessen the challenges associated with small scale farming in Barbados and generally in the Caribbean. The system requires minimal land and water resources, and no soil resources, which is desirable for highly populated and arid regions such as Barbados. Aquaponic systems provide a source of protein as well as fresh fruits, vegetables or herbs. As meat on the island is relatively expensive, protein in the form of freshwater fish would provide a healthy alternative and reduce stress on dwindling saltwater and freshwater fish supplies.

In late 2009 the project "Aquaponic island" was started with joint funding by the United Nations Development Program, the Food and agriculture organization,Ministry of Agriculture, 4H movement, Inter-American Institute for the Cooperation on Agriculture, Bellairs Research Institute And Baird's Village Aquaponic Association

The aim of the project is to educate the island population about the advantages of aquaponics. A commercial size system was set up at Baird's Village and extensive research and development was done adapting the technology to the island climate the next phase is to put aquaponic systems in many of the island schools.

ADVANTAGES

The unique advantages of aquaponic systems are:

- Conservation through constant water reuse and recycling.
- Organic fertilization of plants with natural fish emulsion.
- The elimination of solid waste disposal from intensive aquaculture.
- The reduction of needed cropland to produce like crops.
- The overall reduction of environmental footprint for crop production.
- Small efficient commercial installations can be built close to markets therefore reducing food miles.

DISADVANTAGES

Some conceivable disadvantages with aquaponics are:

- Initial expense for housing, tank, plumbing, pump/s, and grow beds.
- The infinite number of ways in which a system can be configured lends itself to equally varying results, conflicting research, and successes or failures.
- Some Aquaponic installations rely heavily on man-made energy, technology solutions, and environmental control to achieve recirculation and water/ambient temperatures but a system designed with energy conservation in mind can be extremely energy efficient.
- Whilst careful design can minimize the risk, Aquaponics systems can have multiple 'single points of failure' where problems such as an electrical failure or pipe blockage can lead to a complete loss of fish stock.
- Like all aquaculture based systems, stock feed usually consists of fish meal derived from lower value species. Ongoing depletion of wild fish stocks makes this practice unsustainable. There are now,

however, organic fish feeds available which may prove to be a viable alternative and negate this concern. Other alternatives include growing duckweed with an Aquaponics system that feeds the same fish grown on the system, as well as growing Black Soldier Fly larva to feed to the fish using composting grub growers.

MORE USES

Aquaponic systems can be used to replicate controlled wetland conditions that are useful for reclaiming potable water from typical household sewage, in addition to generating a continual supply of food with minimal fertilizer use. Aquaponics takes advantage of synergy between self-organizing biological systems, emphasizing the one element/many functions principle of permaculture as a natural solution for water treatment.

FISH

In practice, tilapia are the most popular fish chosen for home and commercial projects that are intended to raise edible fish. Most green leafy vegetables grow well in the hydroponic filter. Although sometimes selected minerals or nutrients such as iron are added, the main source of nutrients for the plants is the fish waste. In Australia, due to a ban on Tilapia in all states bar W.A., natives are the most popular fish, including Silver Perch, Jade Perch, Sleepy Cod, Murray cod and Barramundi. Rainbow and brown trout while not native to Australia are also in use - along with fresh water crayfish such as yabby and redclaw.

COMPANION PLANTING

Many plants have natural substances in their roots, flowers, leaves etc. that can alternately repel and/or attract insects depending on your needs. In some situations they can also help enhance the growth rate and Flavour of other varieties. Experience shows us that using companion planting through out the landscape is an important part of integrated pest management. In essence companion planting helps bring a balanced eco-system to your landscape, allowing nature to do its' job. Nature integrates a diversity of plants, insects, animals, and other organisms into every ecosystem so there is no waste. The death of one organism can create food for another, meaning symbiotic relationships all around. We consider companion planting to be a holistic concept due to the many intricate levels in which it works with the ecology.

By using companion planting, many gardeners find that they can discourage harmful pests without losing the beneficial allies. There are many varieties of herbs, flowers, etc. that can be used for companion plants. Be open to experimenting and find what works for you. Some possibilities would be using certain plants as a border, backdrop or interplanting in your flower or vegetable beds where you have specific needs. Use plants that are native to your area so the insects you want to attract already know what to look for! Plants with open cup shaped flowers are the most popular with beneficial insects.

Companion planting can combine beauty and purpose to give you an enjoyable, healthy environment. Have fun, let your imagination soar. There are many ways you can find to incorporate these useful plants in your garden, orchard, flower beds etc.

PLANT GUIDE

ALFALFA: Perennial that roots deeply. Fixes the soil with nitrogen, accumulates iron, magnesium, phosphorous and potassium. Withstands droughts with it's long taproot and can improve just about any soil! Alfalfa has the ability to break up hard clay soil and can even send its' roots through rocks! Now that is a tenacious plant! Alfalfa is practically pest and disease free. It needs only natural rainfall to survive.

AMARANTH: A tropical annual that needs hot conditions to flourish. Good with sweet corn, it's leaves provide shade giving the corm a rich, moist root run. Host to predatory ground beetles. Eat the young leaves in salads.

ANISE: Licorice flavored herb, good host for predatory wasps which prey on aphids and it is also said to repel aphids. Deters pests from brassicas by camouflaging their Odour. Improves the Vigour of any plants growing near it. Used in ointments to protect against bug stings and bites. Good to plant with coriander.

ASPARAGUS: *Friends*: Aster family flowers, dill ,coriander, tomatoes, parsley, basil, comfrey and marigolds. Avoid: Onions, garlic and potatoes.

BASIL: Plant with tomatoes to improve growth and Flavour. Basil also does well with peppers, oregano, asparagus and petunias. Basil can be helpful in repelling thrips. It is said to repel flies and mosquitoes. Do not plant near rue or sage.

BAY LEAF: A fresh leaf bay leaf in each storage container of beans or grains will deter weevils and moths. Sprinkle dried leaves with other deterrent herbs in garden as natural insecticide dust. A good combo: Bay leaves, cayenne pepper, tansy and peppermint.

BEANS: All bean enrich the soil with nitrogen fixed form the air. In general they are good company for carrots, celery, chards, corn, eggplant, peas, potatoes, brassicas, beets, radish, strawberry and cucumbers. Beans are great for heavy nitrogen users like corn and grain plants because beans fix nitrogen from the air into the soil so the nitrogen used up by the corn and grains are replaced at the end of the season when the bean plants die back. French Haricot beans, sweet corn and melons are a good combo. Summer savory deters bean beetles and improves growth and Flavour. Keep beans away from the alliums.

BEE BALM: Plant with tomatoes to improve growth and Flavour. Great for attracting beneficials and bees of course. Pretty perennial that tends to get powdery mildew.

BEET: Good for adding minerals to the soil. The leaves are composed of 25% magnesium making them a valuable addition to the compost pile if you don't care to eat them. Beets are also beneficial to beans with the exception of runner beans. Runner or pole beans and beets stunt each other's growth. Companions for beets are lettuce, onions and brassicas. Beets and kohlrabi grow perfectly together. Beets are helped by garlic and mints. Garlic improves growth and Flavour. Rather than planting invasive mints around beets use your mint clippings as a mulch.

BORAGE: Companion plant for tomatoes, squash, strawberries and most plants. Deters tomato hornworms and cabbage worms. One of the best bee and wasp attracting plants. Adds trace minerals to the soil and a good addition the compost pile. The leaves contain vitamin C and are rich in calcium, potassium and mineral salts. Borage may benefit any plant it is growing next to via increasing resistance to pests and disease. It also makes a nice mulch for most plants. Borage and strawberries help each other and strawberry farmers always set a few plants in their beds to enhance the fruits Flavour and yield. Plant near tomatoes to improve growth and disease resistance. After you have planned this annual once it will self seed. Borage flowers are edible.

BRASSICA: Benefit from chamomile, peppermint, dill, sage, and rosemary. They need rich soil with plenty of lime to flourish. Avoid planting with mustards, nightshades.

BUCKWHEAT: Accumulates calcium and can be grown as an excellent cover crop aka green manure. Buckwheat's shallow white blossoms attract beneficial insects that control or parasitize aphids, mites and other pests. The beneficials it attracts include the following: hover flies, predatory wasps, minute pirate bugs, insidious flower bugs, tachinid flies and lady beetles. Flowering may start within three weeks of planting and continue for up to 10 weeks. Buckwheat will take up phosphorus and some minor nutrients that are otherwise unavailable to plants. These nutrients are released as the residue of the buckwheat breaks down and are then available for later crops. The fine roots makes topsoil loose and friable with only minimal tillage.

CABBAGE: Celery, dill, onions and potatoes are good companion plants. Celery improves growth and health. Clover interplanted with cabbage has been shown to reduce the native cabbage aphid and cabbageworm populations by interfering with the colonization of the pests and increasing the number of predatory ground beetles. Plant Chamomile with cabbage as it Improves growth and Flavour. Cabbage does not get along with strawberries, tomatoes, peppers, eggplants, rue, grapes and pole beans.

CARAWAY: Good for loosening compacted soil with it's deep roots so it's also compatible next to shallow rooted crops. Plant it with strawberries. Caraway can be tricky to establish. The flowers attract a number of beneficial insects especially the tiny parasitic wasps. Keep it away from dill and fennel.

CARROTS: Their pals are leaf lettuce, onions and tomatoes. Plant dill and parsnips away from carrots. Flax produces an oil that may protect root vegetables like carrots from some pests. One drawback with tomatoes and carrots: tomato plants can stunt the growth of your carrots but the carrots will still be of good Flavour.

CATNIP: Deters flea beetles, aphids, Japanese beetles, squash bugs, ants and weevils. We have found it repels mice quite well: mice were wreaking havoc in our outbuildings, we spread sprigs of mint throughout and the mice split! Use sprigs of mint anywhere in the house you want deter mice and ants. Smells good and very safe.

CELERY: Companions: Bean, cabbage family, leek, onion, spinach and tomato. Flowers for celery: cosmos, daisies and snapdragons. Foe: Corn.

CHAMOMILE, GERMAN: Annual. Improves Flavour of cabbages, cucumbers and onions. Host to hoverflies and wasps. Accumulates calcium, potassium and sulfur, later returning them to the soil. Increases oil production from herbs. Leave some flowers unpicked and German chamomile will reseed itself. Roman chamomile is a low growing perennial that will tolerate almost any soil conditions. Both like full sun. Growing chamomile of any type is considered a tonic for anything you grow in the garden.

CHARDS: Companions: Bean, cabbage family and onion.

CHERVIL: Companion to radishes, lettuce and broccoli for improved growth and Flavour. Keeps aphids off lettuce. Said to deter slugs. Likes shade.

CHIVES: Improves growth and Flavour of carrots and tomatoes. A friend to apples, carrots, tomatoes, brassica and many others. Keeps aphids help to keep aphids away from tomatoes, mums and sunflowers. Chives may drive away Japanese beetles and carrot rust fly. Planted among apple trees it helps prevent scab and among roses it prevents black spot. You will need patience as it takes about 3 years for plantings of chives to prevent the 2 diseases. A tea of chives may be used on cucumbers and gooseberries to prevent downy and powdery mildews. Avoid planting near beans and peas.

CHRYSANTHEMUMS: C. coccineum kills root nematodes. It's flowers along with those of C. cineraruaefolium have been used as botanical pesticides for centuries. White flowering chrysanthemums repel Japanese beetles. To the right is a picture of the painted daisy from which pyrethrum is extracted.

CLOVER: Long used as a green manure and plant companion and is especially good to plant under grapevines. Attracts many beneficials. Useful planted around apple trees to attract predators of the woolly aphid. Clover interplanted with cabbage has been shown to reduce the native cabbage aphid and cabbageworm populations by interfering with the colonization of the pests and increasing the number of predator ground beetles.

COMFREY: Accumulates calcium, phosphorous and potassium. Likes wet spots to grow in. Comfrey is beneficial to avocado and most other fruit trees. Traditional medicinal plant. Good trap crop for slugs. More on comfrey.

CORIANDER: Repels aphids, spider mites and potato beetle. A tea from this can be used as a spray for spider mites. A partner for anise.

CORN: Amaranth, beans, cucumber, white geranium, lamb's quarters, melons, morning glory, parsley, peanuts, peas, potato, pumpkin, soybeans, squash and sunflower. A classic example is to grow climbing beans up corn while inter-planting pumpkins. The corn provides a natural trellis for the beans, pumpkins smother the weeds and helps corn roots retain moisture. Corn is a heavy feeder and the beans fix nitrogen from air into the soil. The beans do not feed the corn will it is growing but when the bean plants die back they return nitrogen to the soil that was used up by the corn. A win-win situation. Another interesting helper for corn is the weed Pig's Thistle which raises nutrients from the subsoil to where the corn can reach them. Keep corn away from celery and tomato plants.

COSTMARY: This 2-3 foot tall perennial of the chrysanthemum family helps to repel moths.

CUCUMBERS: Cucumbers are great to plant with corn and beans. The three plants like the same conditions warmth, rich soil and plenty of moisture. Let the cucumbers grow up and over your corn plants. A great duet is to plant cukes with sunflowers. The sunflowers provide a strong support for the vines. Cukes also do well with peas, beets, radishes and carrots. Radishes are a good deterrent against cucumber beetles. Dill planted with cucumbers helps by attracting beneficial predators. Nasturtium improves growth and Flavour. Keep sage, potatoes and rue away from cucumbers.

DAHLIAS: These beautiful, tuberous annuals that can have up to dinner plate size flowers repels nematodes!

DILL: Improves growth and health of cabbage. Do not plant near carrots, caraway or tomatoes. Best friend for lettuce. Attracts hoverflies and predatory wasps. Repels aphids and spider mites to some degree. Also may repel the dreaded squash bug! Dill goes well with lettuce, onions, cabbage, sweet corn and cucumbers. Dill does attract the tomato horn worm so it would be useful to plant it somewhere away from your tomato plants to keep the destructive horn worm away from them. Do plant dill in an appropriate spot for the swallowtail butterfly caterpillars to feed on. Even their caterpillars are beautiful.

EGGPLANT: Plant with amaranth, beans, peas, spinach, tarragon, thyme and marigold. Eggplant is a member of the nightshade family and does well with peppers. Avoid planting fennel near eggplant.

ELDERBERRY: A spray made from the leaves can be used against aphids, carrot root fly, cucumber beetles and peach tree borers. Put branches and leaves in mole runs to banish them. Elderberry leaves added to the compost pile speeds up the decomposing process.

FLAX: Plant with carrots, and potatoes. Flax contains tannin and linseed oils which may offend the Colorado potato bug. Flax is an annual from 1-4 feet tall with blue or white flowers that readily self sows.

FOUR-O'CLOCKS: Draws Japanese beetles like a magnet which then dine on the foliage. The foliage is pure poison to them and they won't live to have dessert! It is important to mention that Four O'clock are also poisonous to humans and animals. Be careful where you plant them if you have children and pets. They are a beautiful annual plant growing from 2-3 feet high with a bushy growth form.

GARLIC: Plant near roses to repel aphids. It also benefits apple trees, pear trees, cucumbers, peas, lettuce and celery. Garlic accumulates sulfur: a naturally occurring fungicide which will help in the garden with disease prevention. Garlic is systemic in action as it is taken up the plants through their pores and when garlic tea is used as a soil drench it is also taken up by the plant roots. Has value in offending codling moths, Japanese beetles, root maggots, snails, and carrot root fly. Researchers have observed that time-released garlic capsules planted at the bases of fruit trees actually kept deer away. It's certainly worth a try! Concentrated garlic sprays have been observed to repel and kill whiteflies, aphids and fungus gnats among others with as little as a 6-8% concentration! It is safe for use on orchids too.

GERANIUM: Repels cabbage worms and Japanese beetles, plant around grapes, roses, corn, tomatoes, peppers and cabbage. Geraniums help to distract beet leafhoppers, carrier of the curly top virus.

GOPHER PURGE: Deters gophers, and moles.

GRAPES: Hyssop is beneficial to grapes as are basil, beans, geraniums, oregano, clover, peas, or blackberries. Keep radishes and cabbage away from grapes. Planting clover increases the soil fertility for grapes. Chives with grapes help repel aphids. Plant your vines under Elm or Mulberry trees.

HEMP: Repels many types of beetles which attack brassicas.

HORSERADISH: Plant in containers in the potato patch to keep away Colorado potato bugs. Horseradish increases the disease resistance of potatoes. There are some very effective insect sprays that can be made with the root. Use the bottomless pot method to keep horseradish contained. Also repels Blister beetles. We have observed that the root can yield anti-fungal properties when a tea is made from it.

HOREHOUND: like many varieties in the mint family, the many tiny flowers attract Braconid and Icheumonid wasps, and Tachnid and Syrid flies. The larval forms of these insects parasitize or otherwise consume many other insects pests. It grows where many others fail to thrive and can survive harsh winters. Blooms over a long season, attracting beneficial insects almost as long as you are likely to need them. For best results use horehound directly as a companion plant. Stimulates and aids fruiting in tomatoes and peppers.

HYSSOP: Companion plant to cabbage and grapes, deters cabbage moths and flea beetles. Do not plant near radishes. Hyssop may be the number one preference among bees and some beekeepers rub the hive with it to encourage the bees to keep to their home. It is not as invasive as other members of the mint family making it safer for interplanting.

KELP: When used in a powder mixture or tea as a spray, this versatile sea herb will not only repel insects but feed the vegetables. In particular we have observed that kelp foliar sprays keep aphids and Japanese beetles away when used as a spray every 8 days before and during infestation times. If you have access to seaweed, use it as a mulch to keep slugs away.

KOHLRABI: May be planted with cucumber, onion and chives. Kohlrabi and beets are perfect to grow with one another! Do not plant kohlrabi with pole beans, pepper, strawberry or tomatoes.

LAMIUM: This will repel potato bugs- a big problem for many gardeners!

LARKSPUR: An annual member of the Delphinium family, larkspur will attract Japanese beetles. They dine and die! Larkspur is poisonous to humans too.

LAVENDER: Repels fleas and moths. Prolific flowering lavender nourishes many nectar feeding and beneficial insects. Lavenders can protect nearby plants from insects such as whitefly, and lavender planted under and near fruit trees can deter codling moth. Use dried sprigs of lavender to repel moths. Start plants in winter from cuttings, setting out in spring.

LEEKS: Use leeks near apple trees, carrots, celery and onions which will improve their growth. Leeks also repel carrot flies. Avoid planting near legumes.

LEMON BALM: Sprinkle throughout the garden in an herbal powder mixture to deter many bugs. Lemon balm has citronella compounds that make this work: crush and rub the leaves on your skin to keep mosquitoes away! Use to ward off squash bugs!

LETTUCE: Does well with beets, bush beans, pole beans, cabbage, carrots, cucumbers, onion, radish and strawberries. It grows happily in the shade under young sunflowers.

LOVAGE: Improves Flavour and health of most plants. Good habitat for ground beetles. A large plant, use one planted as a backdrop. Similar to celery in Flavour.

MARIGOLDS: Given a lot of credit as a pest deterrent. Keeps soil free of bad nematodes; supposed to discourage many insects. Plant freely throughout the garden. The marigolds you choose must be a scented variety for them to work.

One down side is that marigolds do attract spider mites and slugs:

- *French Marigold* Has roots that exude a substance which spreads in their immediate vicinity killing nematodes. For nematode control you want to plant dense areas of them. There have been some studies done that proved this nematode killing effect lasted for several years after the plants were These marigolds also help to deter whiteflies when planted around tomatoes and can be used in greenhouses for the same purpose. Whiteflies hate the smell of marigolds. Do not plant French marigolds next to bean plants.

- Mexican marigold is the most powerful of the insect repelling marigolds and may also overwhelm weed roots such as bind weed! It is said to repel the Mexican bean beetle and wild bunnies! Be careful it can have an herbicidal effect on some plants like beans and cabbage.

MARJORAM: As a companion plant it improves the Flavour of vegetables and herbs. Sweet marjoram is the most commonly grown type.

MELONS: Companions: Corn, pumpkin, radish and squash. Other suggested helpers for melons are as follows: Marigold deters beetles, nasturtium deters bugs and beetles. Oregano provides general pest protection.

MINT: Deters white cabbage moths, ants, rodents, flea beetles, fleas, aphids and improves the health of cabbage and tomatoes. Use cuttings as a mulch around members of the brassica family. Mint flowers attract hoverflies and predatory wasps. Earthworms are quite attracted to mint plantings. Be careful where you plant it as mint is an incredibly invasive perennial. We have found that placing peppermint cuttings where mice are a problem is very effective in driving them off!

MOLE PLANTS: Deter moles and mice if planted here and there throughout the garden. Drop a seed of this in mole runs to drive them away. This is a poisonous plant.

MORNING GLORIES: They attract hoverflies. Plus if you want a fast growing annual vine to cover something up morning glory is an excellent choice.

NASTURTIUMS: Nasturtium is an excellent companion for many plants. It is a companion to radishes, cabbage family plants deterring aphids, squash bugs, and striped pumpkin beetles, and improving growth and Flavour. Plant as a barrier around tomatoes, cabbage, cucumbers, and under fruit trees. Deters wooly aphids, whiteflies, cucumber beetles and other pests of the cucurbit family. Great trap crop for aphids which it does attract, especially the yellow flowering varieties. It likes poor soil with low moisture and no fertilizer. Keeping that in mind there is no reason not to set potted nasturtiums among your garden beds. It has been the practice of some fruit growers that planting nasturtiums every year in the root zone of fruit trees allow the trees to take up the pungent Odour of the plants and repel bugs. It is among the best at attracting predatory insects. It has no taste effect on the fruit. A nice variety to grow is Alaska which has attractive green and white variegated leaves. The leaves, flowers and seeds of nasturtiums are all edible and wonderful in salads! Try our recipe for: Nasturtium Salad

NETTLES, STINGING: The flowers attract bees. Sprays made from these are rich in silica and calcium. Invigorating for plants and improves their disease resistance. Leaving the mixture to rot, it then makes an excellent liquid feed. Comfrey improves the liquid feed even more. Hairs on the nettles' leaves contain formic acid which "stings" you.

OKRA: Plant lettuce around your okra plants and they will shade the lettuce in the summer giving you some more growing time. Okra also does well with peppers and eggplants as it helps protect these brittle stemmed plants from high winds. It also gets along with basil, cucumbers, melons, and black eyed peas. For planting with the peas plant your Okra first. When the okra is up and established plant the peas around the edges of the okra planting. You may find that the peas are far less bothered by aphids when near okra.

ONIONS: Planting chamomile and summer savory with onions improves their Flavour. Other companions are carrot, leek, beets, kohlrabi, strawberries, brassicas, dill, lettuce and tomatoes. Intercropping onions and leeks with your carrots confuses the carrot and onion flies! Onions planted with strawberries help the berries fight disease. Keep onions away from peas and asparagus.

OPAL BASIL: An annual herb that is pretty, tasty and said to repel hornworms! Like the other basils it also does well with peppers, oregano, asparagus and petunias. Keep away from rue and sage.

OREGANO: Can be used with most crops but especially good for cabbage. Plant near broccoli, cabbage and cauliflower to repel cabbage butterfly and near cucumbers to repel cucumber beetle. Also benefits grapes.

PARSLEY: Allies: Asparagus, carrot, chives, onions, roses and tomato. Sprinkle the leaves on tomatoes, and asparagus. Use as a tea to ward off asparagus beetles. Attracts hoverflies. Let some go to seed to attract the tiny parasitic wasps and hoverflies.

PEAS: Peas fix nitrogen in the soil. Plant next to corn. Companions for peas are bush beans, Pole Beans, Carrots, Celery, Chicory, Corn Cucumber, Eggplant, Parsley, Early Potato, Radish, Spinach, Strawberry, Sweet pepper and Turnips. Do not plant peas with onions.

PEPPERMINT: Repels white cabbage moths, aphids and flea beetles. It is the menthol content in mints that acts as an insect repellant. Bees and other good guys love it.

PEPPERS, BELL : Plant peppers near tomatoes, parsley, basil, geraniums, marjoram, lovage, petunia and carrots. Onions make an excellent companion plant for peppers. They do quite well with okra as it shelters them and protects the brittle stems from wind. Don't plant them near fennel or kohlrabi. They should also not be grown near apricot trees because a fungus that the pepper is prone to can cause a lot of harm to the apricot tree. Peppers can double as ornamentals, so tuck some into flowerbeds and borders. Harvesting tip: The traditional bell pepper is harvested green, even though most varieties will mature red, orange, or yellow. Peppers can be harvested at any stage of growth, but their Flavour doesn't fully develop until maturity.

PEPPERS, HOT: Chili peppers have root exudates that prevent root rot and other Fusarium diseases. Plant anywhere you have these problems. While you should always plant chili peppers close together, providing shelter from the sun with other plants will help keep them from drying out and provide more humidity. Tomato plants, green peppers, and okra are good protection for them. Teas made from hot peppers can be useful as insect sprays. Hot peppers like to be grouped with cucumbers, eggplant, escarole, tomato, okra, Swiss chard and squash. Herbs to plant near them include: basils, oregano, parsley and rosemary. Never put them next to any beans, broccoli, cabbage, cauliflower, Brussels sprouts or fennel.

PENNYROYAL: Repels fleas. The leaves when crushed and rubbed onto your skin will repel chiggers, flies, gnats, mosquitoes and ticks. Warning: Pennyroyal is highly toxic to cats. It should not be planted where cats might ingest it and never rubbed onto their skin.

PETUNIAS: They repel the asparagus beetle, leafhoppers, certain aphids, tomato worms, Mexican bean beetles and general garden pests. A good companion to tomatoes, but plant everywhere. The leaves can be used in a tea to make a potent bug spray.

POACHED EGG PLANT: Grow poached egg plant with tomatoes, they will attract hover flies and hover flies eat aphids.

POTATO: Companions for potatoes are bush bean, members of the cabbage family, carrot, celery, corn, dead nettle, flax, horseradish, marigold, peas, petunia, onion and Tagetes marigold. Protect them from scab by putting comfrey leaves in with your potato sets at planting time. Horseradish, planted at the corners of the potato patch, provides general protection. Don't plant these around potatoes: asparagus, cucumber, kohlrabi, parsnip, pumpkin, rutabaga, squash family, sunflower, turnip and fennel. Keep potatoes and tomatoes apart as they both can get early and late blight contaminating each other.

PUMPKINS: Pumpkin pals are corn, melon and squash. Marigold deters beetles. Nasturtium deters bugs, beetles. Oregano provides general pest protection.

PURSLANE: This edible weed makes good ground cover in the corn patch. Use the stems, leaves and seeds in stir frys. Pickle the green seed pod for caper substitutes. If purslane is growing in your garden it means you have healthy, fertile soil!

RADISH: One of the workhorses for the garden. Companions for radishes are: radish, beet, bush beans, pole beans, carrots, chervil, cucumber, lettuce, melons, nasturtium, parsnip, peas, spinach and members of the squash family. Why plant radishes with your squash plants? Radishes may protect them from squash borers. Anything that will help keep them away is worth a try. Radishes are a deterrent against cucumber beetles and rust flies. Chervil and nasturtium improve radish growth and Flavour. Planting them around corn and letting them go to seed will also help fight corn borers. Chinese Daikon and Snow Belle radishes are favorites of flea beetles. Plant these at 6 to 12 inch intervals amongst broccoli. In one trial, this measurably reduced damage to broccoli. Radishes will lure leafminers away from spinach. The damage the leafminers do to radish leaves does not stop the radish roots from growing, a win-win situation. Keep radishes away from hyssop plants, cabbage, cauliflower, Brussels sprouts and turnips. For some good eating try our delicious Radish varieties.

RHUBARB: A good companion to all brassicas. Try planting cabbage and broccoli plants your rhubarb patch watch them thrive. Rhubarb protects beans against black fly. Some other interesting companions for rhubarb are the beautiful columbine flowers, garlic, onion and roses! It helps deter red spider mites from the columbines. A spray made from boiled rhubarb leaves, which contain the poison oxalic acid may be used to prevent blackspot on roses and as an aphicide.

ROSEMARY: Companion plant to cabbage, beans, carrots and sage. Deters cabbage moths, bean beetles, and carrot flies. Use cuttings to place by the crowns of carrots for carrot flies. Zones 6 and colder can overwinter rosemary as houseplants or take cuttings.

RUE: Deters aphids, fish moths, flea beetle, onion maggot, slugs, snails, flies and Japanese beetles in roses and raspberries. Companions for rue are roses, fruits, raspberries and lavender. To make it even more effective with Japanese beetles: crush a few leaves to release the smell. Has helped repel cats for us. You should not plant rue near cucumbers, cabbage, basil or sage. A pretty perennial with bluish-gray leaves. May be grown indoors in a sunny window. Rue may cause skin irritation in some individuals.

RYE: An excellent use of plant allelopathy is the use of mow-killed grain rye as a mulch. The allelochemicals that leach from the rye residue prevent weed germination but do not harm transplanted tomatoes, broccoli, or many other vegetables.

SAGE: Use as a companion plant with broccoli, cauliflower, rosemary, cabbage, and carrots to deter cabbage moths, beetles, black flea beetles and carrot flies. Do not plant near cucumbers, onions or rue. Sage repels cabbage moths and black flea beetles. Allowing sage to flower will also attract many beneficial insects and the flowers are pretty. There are some very striking varieties of sage with variegated foliage that can be used for their ornamental as well as practical qualities. More on sage.

SPINACH: Plant with peas and beans as they provide natural shade for the spinach. Gets along with cabbage, cauliflower, celery, eggplant, onion, peas, strawberries.

SOUTHERNWOOD: Plant with cabbage, and here and there in the garden. Wonderful lemony scent when crushed or brushed in passing. Roots easily from cuttings. Does not like fertilizer! It is a perennial that can get quite bushy. We have started to cut it back every spring and it comes back in not time. A delightful plant that is virtually pest free.

SOYBEANS: They add nitrogen to the soil making them a good companion to corn. They repel chinch bugs and Japanese beetles. Why not try soybeans, they are good for you. They are many tasty ways to prepare them.

SQUASH: Companions: Corn, cucumbers, icicle radishes, melon and pumpkin. Helpers: Borage deters worms, improves growth and Flavour. Marigolds deters beetle. Nasturtium deters squash bugs and beetles. Oregano provides general pest protection.

STRAWBERRY: Friends are beans, borage, lettuce, onions, spinach and thyme. Foes: Cabbage, broccoli, Brussels sprouts, cauliflower and kohlrabi. Allies: Borage strengthens resistance to insects and disease. Thyme, as a border, deters worms.

SUMMER SAVORY: Plant with beans and onions to improve growth and Flavour. Discourages cabbage moths, Mexican bean beetles and black aphids. Honey bees love it.

SUNFLOWERS: Planting sunflowers with corn is said by some to increase the yield. Aphids a problem? Definitely plant a few sunflowers here and there in the garden. Step back and watch the ants herd the aphids onto them. We have been doing this for years and it is remarkable. The sunflowers are so tough that the aphids cause very little damage and you will have nice seed heads for the birds to enjoy. Sunflowers also attract hummingbirds which eat whiteflies. Talk about a symbiotic relationship!

SWEET ALYSSUM: Direct seed or set out starts of sweet alyssum near plants that have been attacked by aphids in the past. Alyssum flowers attract hoverflies whose larva devour aphids. Another plus is their blooms draw bees to pollinate early blooming fruit trees. They will reseed freely and make a beautiful groundcover every year.

TANSY: Plant with fruit trees, roses and raspberries keeping in mind that it can be invasive and is not the most attractive of plants. Tansy which is often recommended as an ant repellant may only work on sugar type ants. These are the ones that you see on peonies and marching into the kitchen. At least for us placing tansy clippings by the greenhouse door has kept them out. Deters flying insects, Japanese beetles, striped cucumber beetles, squash bugs, ants and mice! Tie up and hang a bunch of tansy leaves indoors as a fly repellent. Use clippings as a mulch as needed. Don't be afraid to cut the plant up as tansy will bounce back from any abuse heaped on it! It is also a helpful addition to the compost pile with its' high potassium content.

TARRAGON: Plant throughout the garden, not many pests like this one. Recommended to enhance growth and Flavour of vegetables.

THYME: Deters cabbage worms. Wooly thyme makes a wonderful groundcover. You may want to use the upright form of thyme in the garden rather than the groundcover types. Thyme is easy to grow from seeds or cuttings. Older woody plants should be divided in spring.

TOMATOES: Tomato allies are many: asparagus, basil, bean, carrots, celery, chive, cucumber, garlic, head lettuce, marigold, mint, nasturtium, onion, parsley, pepper, marigold, pot marigold and sow thistle. One drawback with tomatoes and carrots: tomato plants can stunt the growth of your carrots but the carrots will still be of good Flavour. Basil repels flies and mosquitoes, improves growth and Flavour. Bee balm, chives and mint improve health and Flavour. Borage deters tomato worm, improves growth and Flavour. Dill, until mature, improves growth and health, mature dill retards tomato growth. Enemies: corn and tomato are attacked by the same worm. Kohlrabi stunts tomato growth. Keep potatoes and tomatoes apart as they both can get early and late blight contaminating each other. Keep cabbage and cauliflower away from them. Don't plant them under walnut trees as they will get walnut wilt: a disease of tomatoes growing underneath walnut trees.

WHITE GERANIUMS: These members of the pelargonum family draw Japanese beetles to feast on the foliage which in turn kills them.

WORMWOOD: Keeps animals out of the garden when planted as a border. An excellent deterrent to most insects. Don't plant wormwood with peas or beans. A tea made from wormwood will repel cabbage moths, slugs, snails, black flea beetles and fleas effectively. The two best varieties for making insect spray are Silver King and Powis Castle. Adversely Powis castle attracts ladybugs which in turn breed directly on the plant. Silver Mound is great as a border plant and the most toxic wormwood. Note: As wormwood actually produces a botanical poison do not use it directly on food crops.

YARROW: Yarrow has insect repelling qualities and is an excellent natural fertilizer. A handful of yarrow leaves added to the compost pile really speeds things up. Try it! It also attracts predatory wasps and ladybugs to name just two. It may increase the essential oil content of herbs when planted among them. Yarrow has so many wonderful properties to it and is an ingredient in our own Golden Harvest Fertilizer.

ZINNIA: Pretty zinnias attract hummingbirds which eat whiteflies. Alternately the pastel varieties of zinnias can be used as a trap crop for Japanese beetles. All zinnias attract bees and other insect pollinators.

INTERCROPPING

Intercropping offers farmers the opportunity to engage nature's principle of diversity on their farms. Spatial arrangements of plants, planting rates, and maturity dates must be considered when planning intercrops. Intercrops can be more productive than growing pure stands. Many different intercrop systems are discussed, including mixed intercropping, strip cropping, and traditional intercropping arrangements. Pest management benefits can also be realized from intercropping due to increased diversity. Harvesting options for intercrops include hand harvest, machine harvest for on-farm feed, and animal harvest of the standing crop.

PRINCIPLES

Sustainable agriculture seeks, at least in principle, to use nature as the model for designing agricultural systems. Since nature consistently integrates her plants and animals into a diverse landscape, a major tenet of sustainable agriculture is to create and maintain diversity. Nature is also efficient. There are no waste products in nature. Outputs from one organism become inputs for another. One organism dies and becomes food for other organisms. Since we are modeling nature, let us first look at some of the principles by which nature functions. By understanding these principles we can use them to reduce costs and increase profitability, while at the same time sustaining our land resource base.

Diversity is Nature's Design

When early humans replaced hunting and gathering of food with domestication of crops and animals, the landscape changed accordingly. By producing a limited selection of crop plants and animals, humankind has greatly reduced the level of biological diversity over much of the earth. Annual crop monocultures represent a classic example. In response to this biological simplification, nature has struggled to restore diversity to these landscapes—that is her tendency. Our "war" with nature over the tendency to diversity is what we call "weed control" and "pest management." Of course we could hardly produce any crops if we simply allowed our fields to return to natural vegetation, but we can realize some of the benefits of diversity by planting mixtures of different crops.

Cooperation is More Apparent than Competition

There is far more cooperation in nature than competition. Cooperation is typified by mutually beneficial relationships that occur between species within communities. In *The Redesigned Forest*, ecologist Chris Maser offers a glimpse of the cooperation inherent in a northern temperate forest when he describes a relationship that exists among squirrels, fungi, and trees. The squirrels feed on the fungus, then assist in its reproduction by dropping fecal pellets containing viable fungal spores onto the forest floor. There new fungal colonies establish. Tree feeder roots search out the fungi and form a symbiotic association that enables the tree roots to increase their nutrient uptake. The fungi, in turn, derive food from the tree roots. Each benefits from the other's presence or actions.

If we view competition as the driving force in nature, we fail to see the complex relationships and feel compelled to take actions that may have unforeseen impacts. The rancher who views coyotes as competitors and kills them out may later find the predator helped keep rodent populations in check. With the predator gone, rodent numbers explode and cause more problems than ever before. The same is true with many insect pests of crops. When the only food for insects is crops, that is what they will eat. With no predator or parasite habitat present in a pure stand of crop, the pest insect could not possibly have it better. If we can shift our view of nature from a theme of competition to one of collaboration, we can act in ways that yield fewer negative consequences.

Stability Tends to Increase with Increasing Diversity

If left undisturbed and unplanted, an abandoned crop field will first be colonized by just a few species of plants, insects, bacteria, and fungi. After several years, a complex community made up of many wild species develops. Once a wild plant and animal community has reached a high level of diversity, it remains stable for many years.

When wild communities are in the early stages of development, or when they have lost diversity due to natural catastrophe or human actions, they are prone to major fluctuations, both in types of species present and in their numbers. Disease outbreaks in plants and animals occur more frequently—as do outbreaks of weed, insect, bird, or rodent pests. One good example is the grasshopper plagues that follow regional weather shifts. Another is the shift in weed species dominance following a soil disturbance.

The more complex and diverse communities become, the fewer the fluctuations in numbers of a given species, and the more stable the communities tend to be. As the number of species increases, so does the web of interdependencies. In both higher and lower rainfall years, there are fewer increases in any one species and fewer fluctuations in the community as a whole.

PURSUING DIVERSITY ON THE FARM

So, then, how can we begin to model our agricultural pursuits after some of these natural principles? Can we look for patterns in nature and imitate them? Some pioneering farmers have been able to utilize nature's principle of diversity to their advantage. Results of their efforts include lower cost of production and higher profits. Among the practices that promote diversity and stability are:

Enterprise diversification: Risk reduction through stability of income and yield are two of the reasons people diversify their crop and livestock systems. Increasing diversity on-farm also reduces costs of pest control and fertilizer, because these costs can be spread out over several crop or animal enterprises.

Crop Rotation: Moving from simple monoculture to a higher level of diversity begins with viable crop rotations, which break weed and pest life cycles and provide complementary fertilization to crops in sequence with each other.

Farmscaping: Diversity can be increased by providing more habitat for beneficial organisms, habitats such as borders, windbreaks, and special plantings for natural enemies of pests.

Intercropping: Intercropping is the growing of two or more crops in proximity to promote interaction between them. Much of this publication focuses on the principles and strategies of intercropping field crops. A related ATTRA publication, *Companion Planting*, provides more information on intercropping of vegetable crops.

Integration: On-farm diversity can be carried to an even higher level by integrating animals with intercropping. With each increase in the level of diversity comes an increase in stability. This publication focuses on intercropping and provides a section on integrating livestock with crops.

INTERCROPPING CONCEPTS

Most grain-crop mixtures with similar ripening times cannot be machine-harvested to produce a marketable commodity, since few buyers purchase mixed grains. Because of limited harvest options with that type of intercropping, farmers are left with the options of hand harvesting, grazing crops in the field with animals, or harvesting the mixture for on-farm animal feed. However, some intercropping plans allow for staggered harvest dates that keep crop species separated. One example would be harvesting wheat that has been interplanted with soybeans, which are harvested later in the season. Another example is planting harvestable strips, also known as strip cropping.

When two or more crops are growing together, each must have adequate space to maximize cooperation and minimize competition between them.

To accomplish this, four things need to be considered:

- Spatial arrangement,
- Plant density,
- Maturity dates of the crops being grown, and
- Plant architecture.

Spatial Arrangement

There are at least four basic spatial arrangements used in intercropping.

Most practical systems are variations of these:

- *Row intercropping*: Growing two or more crops at the same time with at least one crop planted in rows.
- *Strip intercropping*: Growing two or more crops together in strips wide enough to permit separate crop production using machines but close enough for the crops to interact.
- *Mixed intercropping*: Growing two or more crops together in no distinct row arrangement.
- *Relay intercropping*: Planting a second crop into a standing crop at a time when the standing crop is at its reproductive stage but before harvesting.

Plant Density

To optimize plant density, the seeding rate of each crop in the mixture is adjusted below its full rate. If full rates of each crop were planted, neither would yield well because of intense overcrowding. By reducing the seeding rates of each, the crops have a chance to yield well within the mixture. The challenge comes in knowing how much to reduce the seeding rates. If you are planning to grow corn and cowpeas, and you want mostly peas and only a little corn, it would be easy to achieve this. The corn-seeding rate would be drastically cut (by 80% or more), and the pea rate would be near normal. The field should produce near top yields of peas even from the lower planting rate and offer the advantage of corn plants for the pea vines to run on. If you wanted equal yields from both peas and corn, then the seeding rates would be adjusted to produce those equal yields.

Maturity Dates

Planting intercrops that feature staggered maturity dates or development periods takes advantage of variations in peak resource demands for nutrients, water, and sunlight. Having one crop mature before its companion crop lessens the competition between the two crops. An aggressive climbing bean may pull down corn or sorghum growing with it and lower the grain yield. Timing the planting of the aggressive bean may fix the problem if the corn can be harvested before the bean begins to climb. A common practice in the old southern U.S. cotton culture was to plant velvet beans or cowpeas into standing corn at last corn cultivation. The corn was planted on wide 40-inch rows at a low plant population, allowing enough sunlight to reach the peas or beans. The corn was close enough to maturity that the young legumes did not compete. When the corn was mature, the beans or peas had corn stalks to climb on. The end result was corn and beans that would be hand harvested together in the fall. Following corn and pea harvest, cattle and hogs would be turned into the field to consume the crop fodder. Selecting crops or varieties with different maturity dates can also assist staggered harvesting and separation of grain commodities. In the traditional sorghum/pigeonpea intercrop, common in India, the sorghum dominates the early stages of growth and matures in about four months. Following harvest of the sorghum, the pigeonpea flowers and ripens. The slow-growing pigeonpea has virtually no effect on the sorghum yield.

Plant Architecture

Plant architecture is a commonly used strategy to allow one member of the mix to capture sunlight that would not otherwise be available to the others. Widely spaced corn plants growing above an understory of beans and pumpkins is a classic example.

INTERCROP PRODUCTIVITY

One of the most important reasons to grow two or more crops together is the increase in productivity per unit of land. Researchers have designed a method for assessing intercrop performance as compared to pure stand yields. In research trials, they grow mixtures and pure stands in separate plots. Yields from the pure stands, and from each separate crop from within the mixture, are measured.

From these yields, an assessment of the land requirements per unit of yield can be determined. This information tells them the yield advantage the intercrop has over the pure stand, if any. They then know how much additional yield is required in the pure stand to equal the amount of yield achieved in the intercrop. To calculate an LER, the intercrop yields are divided by the pure stand yields for each component crop in the intercrop. Here's the equation for a corn/pea intercrop where the yields from pure corn, pure peas, and the yields from both corn and peas growing together in an intercrop are measured.

(intercrop corn / pure corn) + (intercrop pea / pure pea) = LER

When an LER measures 1.0, it tells us that the amount of land required for peas and corn grown together is the same as that for peas and corn grown in pure stand. LERs above 1.0 show an advantage to intercropping, while numbers below 1.0 show a disadvantage to intercropping. An LER of 1.25 tells us that the yield produced in the total intercrop would have required 25% more land if planted in pure stands. If the LER was 0.75, then we know the intercrop yield was only 75% of that of the same amount of land that grew pure stands.

In a South Carolina study, researchers planted intercrops of southern peas and sweet corn at three different corn plant densities. The plantings were on raised beds with flat and wide crowns on six-foot centers. In the center of each bed was a corn row, with two rows of peas planted 18 inches to either side of the corn row. The low corn-seeding rate was 6,700 plants per acre, medium corn was 9,500 per acre, and high was 11,900 plants per acre. Peas were established at a rate of 31,800 plants per acre in all intercrop plots. In the pure pea stand, each bed had two rows of peas spaced 24 inches apart. Yields of the intercrops and pure stands are shown in Table.

In this trial there was a yield advantage from intercropping over growing the two crops in pure stands. Pea yields suffered from the increased competition in the higher densities of corn. Some practical on-farm guidelines can be drawn to guide seeding-rate choices for a two-crop intercrop. To test seeding rates, experiment with three small plantings of two crops at the following percentages of their full seeding rates: 1/3 + 2/3, 1/2 + 1/2, and 2/3 + 1/3. From there, make adjustments for future plantings based on the results and your expectations.

Table. Yields of sweet corn and southern peas from intercrops.

Seed Rates	Corn (pounds/acre)	Peas	LER (pounds/acre)
Full corn	5600	***	***

Full peas	***	1200	***
Low corn	4200	800	1.41
Medium corn	4600	800	1.48
High corn	5000	500	1.30

MANAGING INTERCROPS

Many combinations of crops have been grown or experimented with as mixed or relay intercrops. Some of these include sunflowers grown with black lentils, wheat with flax, and canola with flax. Other combinations include cucumbers, beans, celery, and chives in China; upland rice, corn, and cassava in Indonesia, and in various parts of the tropics corn and cassava, corn and peanut, sorghum and millet, and sorghum and pigeonpeas.

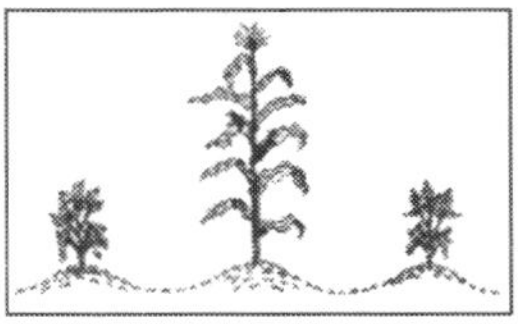

Fig. Sweetcorn and Southern Pea Planting Pattern.

Frequently these cropping combinations involve a short and a tall crop both planted at the same time. In many cases the tall crop is harvested first. Corn grown with a shorter plant would be harvested first, then peanut or sweet potato would be harvested later. Another pattern would be planting two tall crops with different growth rates. In relay intercrops, different planting dates are used so that one crop might mature sooner. Corn or sorghum, requiring three months to mature, can be grown with pigeonpea, requiring 10 months to maturation.

John Bowen and Bernard Kratky, researchers and instructors at the University of Hawaii, tell us that there are five distinct aspects to successful multiple cropping.

These are:

- Detailed planning,
- Timely planting of each crop,
- Adequate fertilization at the optimal times,
- Effective weed and pest control,
- Efficient harvesting.

Before any fieldwork is begun, adequate planning should be done. Planning covers selection of crop species and appropriate cultivars, water availability, plant populations and spacing, labor requirements throughout the season, tillage requirements, and predicted profitability of the intercrop. These and other parameters need to be evaluated before spending money on inputs. With any crop, seed germination and seedling establishment are the most critical phases of the entire season. A good seedbed is needed to get a good stand. Delayed planting may reduce yield, since crop development may not coincide with the optimal growth periods.Planning fertilization for intercrops can be challenging, as the full needs of both crops must be met. Generally, there is little information available on how to go about this. One possibility would be to ask for soil test results for each crop separately, then formulate a recommendation that will cover the needs of both crops to be grown. Such recommendations are generally 10% to 30% higher than rates for individual crops.

As with any crop, also accounting for residual or carryover fertility from past crops saves money. Carryover fertility from intercrops may well be lower than that of pure stands because of the two crops having different root types and feeding habits.Weed and pest controls need in intercrops will likely be different from those in pure stands. Some disease incidence, such as soybean or mung bean rusts, may increase when aggravated with high corn populations and overfertilization. Any disease or pest that prospers in shady conditions could increase under a taller crop such as corn or sunflowers. In many cases, insect pest populations are lower when two or more crops are grown together. More on pest management will be found later in this publication.

Harvesting of mixed intercrops has been a major limitation to their adoption in mechanized farming. If the crops cannot be harvested by animals, or all together as feed, you're left with hand harvesting. Some crops such as flax and wheat have been harvested together and mechanically separated. Any other mechanized harvest efforts must get one crop without damaging the other. One example would be harvesting wheat over the top of a young stand of soybeans growing beneath the grain heads. All intercropping strategies— especially mixed intercropping—require advanced planning and keen management. Success will likely be the reward for such efforts.

EXAMPLES OF INTERCROP SYSTEMS

Traditional Corn-Bean-Squash Mixed Intercrops

Farmers throughout Central America traditionally grow an intercrop of corn, beans and squash. Grown together, these three crops optimize available resources. The corn towers high over the other two crops, and the beans climb up the corn stalks. The squash plants sprawl along the ground, capturing light that filters down through the canopy and shading the ground. The shading discourages weeds from growing. This mixture was compared to the individual crops grown separately in a study near Tabasco, Mexico. In the study, corn yields were considerably higher in the mixture than in a pure stand planted at optimum densities. Bean and squash yields suffered considerable yield reductions when grown in mixture. In this example if corn were the most important crop, it was beneficial to grow it in a mixture with squash and beans. The beans and squash were just a bonus. The LER for the whole mixture was considerably higher than any of the pure stands.

Table. Yields of corn, beans and suash grown alone or in a mixture

Crop	Pure Stand	Intercrop (pounds/acre)(pounds/acre)
Corn	1096	1533
Beans	544	98
Squash	383	71

Corn and Soybean Mixed Intercrops

Canadian researchers have worked with several corn-soybean intercrop seeding rates to determine their economic advantages as silage. Pure stands of corn and soybeans were grown for comparison at 24,000 corn seed per acre and 200,000 soybean seed per acre. Results showed that intercrops were more cost effective than pure stands over both years the study was conducted. The study featured five experimental intercrop seeding rates with two planting arrangements The researchers concluded that a planting rate of 16,000 corn seed per acre with 135,000 soybean seed per acre planted within the same rows along with 53 lbs. of N/acre gave the highest economic returns. This mixture gave an LER of 1.14 over pure stand yields. The crude protein level of the intercrop silage was considerably higher than that of pure corn silage. A slightly higher yield was achieved from full stands of both corn and beans in alternate rows, but the cost of production was higher, thus offsetting the improved yields.

Corn and Sorghum Mixed Intercrops

Frank Cawrse, Jr., of Lebanon, Oregon, intercrops forage sorghum into his silage corn. He first plants the corn at 28,000 seed per acre, then goes back over the field with a drill with enough drop tubes closed off to plant 8 pounds of sorghum on 32-inch rows in between the corn. He also plants two different maturities of corn, a 95-day and a 75-day, to even out the silage moisture content. He harvests a mix of corn in hard dent and soft dent, and sorghum in the milk stage.

Strip Cropping Corn/Soybeans/Small Grains

South Dakota farmer Tod Intermill plants alternating strips of corn, soybeans, and spring wheat on his farm. The strips are six rows wide in a ridge-till system. All the crop plantings are adapted to existing equipment widths. Regular herbicide treatments can be applied using a ground sprayer of strip width. Even the wheat is drilled on ridges, using a drill with individual depth gauges on each opener. Intermill orients his rows east and west to minimize the shading effects of taller crops like corn. The crops are planted in a wheat-corn-soybean pattern, with soybeans on the north side of the corn. This arrangement reduces the effect of corn shading often associated with a straight corn-soybean pattern, since the wheat is mature before the corn has a chance to shade it. Corn gains the greatest benefit from the additional sunlight interception on the outside rows of the corn strip.Iowa farmer Tom Frantzen strip-crops oats, corn, and soybeans on ridge-till rows. He views his strips as a crop rotation in one field. His rows are oriented generally east and west on the contour.

His 1989 strip-crop corn yields were 166 bushels per acre, compared to 130 for his farm average. Stripped soybean yields were two bushels lower than farm average. His oat yields were 109 bushels stripped and 100-bushel farm average. Tom was not surprised at the increase in corn yields. The outer strip rows captured more sunlight. His average corn border row yielded 198 bushels per acre next to the soybeans and 177 bushels next to oats. The soybean yields were 37 bushels, even with the increased shading on the border rows. This loss was made up in the middle rows with yields of 44 bushels per acre. Oats showed a 107-bushel yield on the soybean side, a 103-bushel yield on the corn side, and 99 bushels in the middle. Tom says the strip intercropping is no more Labour intensive than monocrop fields. His profits were $76 per acre for the stripped fields and $55 for the same crops grown in monoculture.

Rick Cruse, an Iowa State University agronomist, has observed several characteristics that narrow strips offer. The strips accommodate the pest management and soil building advantages of rotations and the yield boost of border rows. With proper management the border effect can pay off; managed improperly, it can cost yield. With oat and corn strips, the early-maturing oats are nearly mature before corn can pose much of a shade and competition problem. The corn can also provide wind protection for the oats. When the oats are harvested, more sunlight is available to the corn. In times of low moisture, oats may rob the corn border rows of water. In his field trials, Cruse found a 5% increase in oat yields on their borders while corn realized a 12 to 15% increase.

Soybean yields dropped by 10% on their border rows, but the yields in the soybean middle rows were higher than they would be in a solid field, possibly representing a windbreak effect.

Some have experimented with a shorter corn variety in the border row to minimize shading. One farmer tried planting six rows of corn and doubling his soybean strips to 12 rows to eliminate the impact of corn shading on the beans. This same farmer found that corn strips wider than eight rows did not provide adequate results. Using a 12-row planter, it's easy to establish the 6-row strips by filling the middle six hoppers with corn and the outer three hoppers with beans. Some farmers plant higher corn populations and add higher nitrogen rates in the border rows to take advantage of the extra sunlight exposure. Most farmers agree that strip cropping corn, soybeans, and oats works best with ridge-till or no-till. When the field is tilled, it's difficult to gauge where the rows should go in order to get the strips even.

ESCALATING DIVERSITY AND STABILITY TO A HIGHER LEVEL

The diversity created by intercropping can be enhanced even further by integrating livestock (single or mixed species) into the cropping plan as harvesters. Allowing animals to harvest feed crops in the field puts gain on animals at the cost of crop production—considerably less than the purchase price of the grain. If you think about it, feed grains cost a lot less when they're not run through a $150,000 combine or hauled 1000 miles across the country.

Grazing animals and other livestock can be managed on croplands to reduce costs, increase income, and increase diversity. There are ways of incorporating animals into cropping without the farmer getting into animal husbandry or ownership directly. Collaboration with neighbors who own animals will benefit both croppers and livestock owners. Grazing or hogging-off of corn residue is one example where a cost can be turned into a profit. The animals replace the $6 per acre stalk mowing cost and produce income in animal gains.

Shasta College provides a unique demonstration of integrating livestock with intercrops. Shasta is a two-year community college located in Redding, California, that offers associate degrees in several branches of agriculture. Stan Gorden heads the college's holistic resource laboratory, where students get hands-on experience with ranching and farming. Stan and his students have taken intercropping to a high level of efficiency. They run hogs, sheep, cattle, and chickens together over 42 small paddocks of various forages and crops growing on 100 acres of college-owned land. One paddock is a pumpkin patch, another a garlic and carrot patch. Some are planted in alfalfa or mixes of grasses and clover. Not all the pastures have water sources for the animals, so water is moved on a trailer tank when necessary. The animals are moved daily in a planned grazing system during rapid plant growth and much more slowly, up to five days on a paddock, during slow plant growth.

Some of the paddocks are planted with mixtures of either winter or summer forage or grain crops. An intercrop of cereal grain, fava beans, and Canadian field peas is planted for winter grain, each crop at 1/3 normal seeding rate. The grain mixture is combine-harvested to make energy and protein supplement feed as needed. After harvest, the animals are turned into the paddock to glean what's left. For summer feed, a mixture of milo planted on 18-inch rows is intercropped with a row of black-eyed peas planted six inches to either side of each sorghum row, using a drill with partitions in the seedbox. The milo provides a trellis for the pea vines to run on. The milo/black-eyed mixture requires no herbicide. Before peas and milo were grown together, the milo pure stand would be plagued with whiteflies and green bugs. Mixing the two crops together ended the pest problem. Cowpeas have extrafloral nectaries that attract lots of beneficial insects. This could explain the absence of pest insects in the mixture. The milo/pea mixture is harvested by setting the combine to cut at the height of the milo heads. This yields a milo to bean ratio of 2:1—ideal for feed. The college animal herd consists of 20 sows that farrow on pasture, 35 head of cattle, 50 sheep, and 30 laying hens that all range together. The hens are with the herd during the day and roost in a nearby eggmobile at night. Gorden selects breeds and genetics to fit this system, as opposed to selecting breeds for maximum production, and adapting a system to match the animal.

The animals benefit one another. The sheep learn to stay close to the middle of the herd to avoid predators, which are fended off by the hogs. The cattle learn that the hogs know how to break the pumpkins open, so they stick close and get some too. The hogs eat the cow and sheep droppings and benefit from the predigestion. The hens scavenge wasted seeds from the various crops. There are three different kinds of hens, each of which lays eggs of a different color. The eggs are marketed as rainbow eggs, with each dozen containing four white, four blue, and four brown eggs. The chickens also scratch apart cattle dung pats searching for insects, thus destroying cattle parasites.

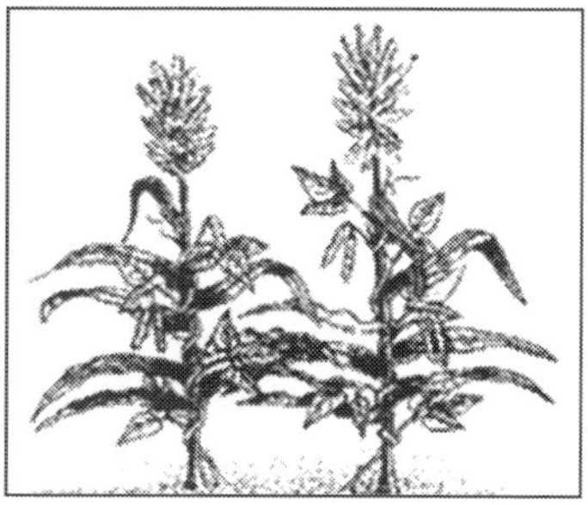

Fig. Cowpeas and Milo Growing Together.

Gorden, that developing and maintaining this high level of diversity has required creativity, selection criteria, constant monitoring, and re-examining traditional beliefs. By challenging long-held beliefs, he and his students discovered that hogs do not need standard farrowing crates and that sheep and cattle are compatible grazers. Animals and crops are selected and culled according to their ability to adapt to this complex system. Shasta College has one of the largest heritage hog herds in the country.

The hogs have been fitted with humane nose rings to prevent rooting. Also, hog breeds are selected that don't root up the ground nor eat the baby lambs when they are born. The sows farrow on pasture with only a single bale of hay for bedding. Hogs are not vaccinated, nor are needle teeth removed or other detailing done. Sows generally wean 12 pigs with no supplemental feed. The only purchased input is some nitrogen and phosphorus fertilizer applied to the pastures. The pigs are only touched twice; once to castrate and once to wean. As with the hogs, the cattle and sheep are selected to prosper on grass. Predators are not controlled in any way. Any animal that gets killed by wandering off is naturally selected out of the herd. The sheep/hog/cow mix provides much better utilization of forage than single species grazing. Since the animals do most of the harvesting, less fossil fuel and Labour-hours are expended. There are no pens to wash and no manure to deal with. The herd is controlled using an electric fence charged up to 8,000 volts to hold the sheep.

Before the 100-acre crop/animal integration project began in 1987, the College's agriculture resource laboratory was costing $8,000 per year. That was the first year the resource laboratory started managing holistically. By 1996, the resource lab's income was up $12,000, and expenses were down $10,000—rendering a $14,000 profit over the 1987 figure. During that same time the soil organic matter has increased from 1.7% to 3.2 per cent.

INTERCROPPING FOR DISEASE CONTROL

Under direction of an international team of scientists, farmers in China's Yunnan province made some simple changes in their rice production methods. They changed from planting their typical pure stand of a single rice variety to planting a mixture of two different rice varieties. Their primary reason for trying this new technique was to reduce the incidence of rice blast, the main disease of rice. The technique was so successful at reducing blast disease that the farmers were able to abandon chemical fungicides they had been using.

The biodiversity effect is apparent here in that if one variety of a crop is susceptible to a disease, the denser the stand, the worse the disease can spread. If susceptible plants are separated by non-host plants that can act as a physical barrier to the disease, the susceptible variety will suffer less disease infection. Rice blast moves from plant to plant via airborne spores. These spores can be blocked by a row of a resistant variety. In this on-farm study, the rice was harvested by hand. Separating the varieties was easily done during harvest, since one variety towered above the other.

ADAPTING INTERCROPPING TO YOUR FARM

Intercropping has been important in the U.S. and other countries and continues to be an important practice in developing nations. In traditional systems, intercropping evolved through many centuries of trial and error. To have persisted, intercropping had to have merit biologically, environmentally, economically, and sociologically. To gain acceptance, any agricultural practice must provide advantages over other available options in the eyes of the practitioner . Many of the impediments to adoption of new strategies or practices of diversification are sociological and financial rather than technological.

Farmers have generally regarded intercropping as a technique that reduces risks in crop production; if one member of an intercrop fails, the other survives and compensates in yield to some extent, allowing the farmer an acceptable harvest. Pest levels are often lowered in intercrops, as the diversity of plants hampers movement of certain pest insects and in some cases encourages beneficial insect populations.

MULTIPLE CROPPING

Planting two or more species in the same field in the same year. Preserved through history to maintain biological, economic, and nutritional diversity, multiple-species systems still are used by the majority of the world's farmers, especially in developing countries. Where farm size is small and the lack of capital has made it difficult to mechanize and expand, farm families that need a low-risk source of food and income often use multiple cropping. These systems maintain a green and growing crop canopy over the soil through much of the year, the total season depending on rainfall and temperature. Systems with more than one crop frequently make better use of total sunlight, water, and available nutrients than is possible with a single crop. The family has a more diverse supply of food and more than one source of income, with both spread over much of the year.

Multiple-cropping patterns are described by the number of crops per year and the intensity of crop overlap. Double cropping or triple cropping signifies systems with two or three crops planted sequentially with no overlap in growth cycle. Intercropping indicates that two or more crops are planted at the same time, or at least planted so that significant parts of their growth cycles overlap. Relay cropping describes the planting of a second crop after the first crop has flowered; in this system there still may be some competition for water or nutrients. When a crop is harvested and allowed to regrow from the crowns or root systems, the term ratoon cropping is used. Sugarcane, alfalfa, and sudangrass are commonly produced in this way, while the potential exists for such tropical cereals as sorghum and rice. Mixed cropping, strip cropping, associated cropping, and alternative cropping represent variations of these systems.

NO-DIG GARDENING

No-Dig Gardening is such a brilliant form of home-based agriculture. The TreeHugger archives would be rich with its merits. Was very surprised when one only found one mention, in a post chronicling Leonora's permaculture adventures in New Zealand. So one launched into the following first-person account of No-Dig, only to discover that in North America the same process might be better known as as Sheet Mulching. Nomenclature aside, it's worth covering the topic again. Especially if you want to grow your own veggies for a little food security.

BACKGROUND TO CULTIVATION FREE FARMING

No-Dig Gardening can probably trace its legacy back to visionary Japanese agricultural pioneer, Fukuoka Masanobu, who embarked on his Natural Farming experiment in 1938. His very productive farming methods did not require extensive soil tilling, weeding, or application of synthetic pesticide or fertilizer. Fukuoka Masanobu advocated returning grain and rice straw stalks to the fields as a way of enriching soil development.

Ruth, though maybe lacking some of the quiet humility and philosophy of her Japanese predecessor, also promoted covering gardens in a dense layer of straw and green mulch.

In the Antipodes we had Esther Dean, who released her own book *Growing Without Digging* in 1977, seeding a small cult following of No Dig gardeners. And, of course, Bill Mollison and David Holmgren who refined their concept of a nature-inspired agriculture with the publication of Permaculture One in 1978.

All would champion the idea that soil quality will dramatically improve if left undisturbed by cultivating, tilling, plowing, digging etc. They believed that soil was enriched with top layers of mulch decomposing to develop the appropriate communities of worms and micro-organisms that enhance food growth. Their ideas have since been embraced even in broad acre agriculture under the guise of no-till farming.

SO HOW DOES IT WORK IN PRACTICE?

There are many ways to implement a no-dig garden.

What follows is just one method:

- We selected a section of the yard that would get at least six hours of direct sunlight. Unfortunately we had to cut down a couple of trees to ensure this access when the sun dropped to lower plane in winter.
- We set up four main beds, so we can practice crop rotation, which rests the soil and reduces chances of plant pests making a comfortable home in the soil.

  Your first bed might get the root crops such as carrots, onions, beetroot and potatoes. The second is for Curcurbits, which are melons, pumpkins, squash, zucchini and cucumbers. Corn can also be planted here. For the third bed consider Acid Lovers: tomatoes, chillies, capsicums and eggplant. And in the last one go the Legumes, like peas and beans and the Brassicas. Each year plant the same vegetables per bed, but one bed further round the rotation.

  Separate non-rotating beds for herbs, and for perennials such as asparagus, strawberries and rhubarb, can also benefit from no dig methods.
- Because our soil is exceedingly clayey, we sprinkled gypsum through the grass to help loosen up the clay. Then we laid out salvaged railway sleepers to give the four main garden beds some definition. These beds were well watered down.

- Atop the wet grass we laid out big sheets of cardboard. This helps supress weeds. The cardboard was thoroughly wetted down too.
- A bale of chopped lucerne straw was spread over the soggy cardboard. And half a bale of long stem lucerne straw covered the lighter chop. This was watered in also.
- On this went a thick layer of what might be called soil 'matter.' It

comprised material rescued from the floor of an old chicken coop that we'd shovelled out, and sieved through an old wire bed frame. It was a mixture of ancient chicken manure, soil, sawdust and compost scraps. Wetted down too.

- Over this soil matter we scattered a bale and half of plain straw and gave the whole shebang a deep soaking.
- We let it 'stew' a little while we waited for our organic seedling and seeds to arrive in the mail. And for the truck to drop off the the 2 tonnes of garden soil, mixed with cow manure.
- With these other ingredients in place we spread apart, with a trowel, strategic holes in the decomposing layered no-dig garden bed. Into these holes we dropped a couple of scoops of the soil/manure. Using a scoop home-made from a juice container. The beauty of this approach is that you only need use soil where you have plants. It's cheaper and saves on the volume of soil that needs to be humped about in a wheel barrow.

- Using a 'dibbler' stick we created a hole in the soil, and inserted the seedlings and seeds to their recommended depth and spacing. These were watered in with mixture of water and seaweed extract, to promote root growth. Then we pulled the straw loosely back over the mini-plots to reduce the soil from drying out.
- Of course it doesn't take long for snails and slugs to find these succulent new growths. So we cut up more containers to make ramps on the outside and shallow dished inside, which we filled with beer

or wine. The slugs tempted by the sweet aroma slide up the ramp and succumb to alcohol poisoning. We also sprinkle some pine needles around the plants to create a spikey surface over which they need to crawl. Though this latter measure can make the soil a bit acidic, so we might experiment with wood ash from the slow combustion heater instead. We've also gone out at night with a head torch to pick off the sneaky slimy critters who don't fall for the beer traps.

- Water for the first couple of weeks to help the seeds and seedlings establish. Then let the straw mulch provide the soil with shade and to hold in any moisture from rain, dew or fog. But otherwise the garden should look after itself by-and-large. If weeds do poke through they can be pulled up, or simply smothered with another layer of straw.

The many steps indicated here might make it seem a drawn out process. But if you had everything together, it could all be set up one day on the weekend. Once set up, your No-Dig Garden shouldn't need more than a few hours tending each week. And besides it is far more satisfying to walk out into the backyard to harvest your own food, than to bump and grind around supermarket aisles and carparks. Cheaper, healthier and saves on the gym subscription too.

POLLINATION MANAGEMENT

Pollination Management is the label for horticultural practices that accomplish or enhance pollination of a crop, to improve yield or quality, by understanding of the particular crop's pollination needs, and by knowledgeable management of pollenizers, pollinators, and pollination conditions.

POLLINATOR DECLINE

With the decline of both wild and domestic pollinator populations, pollination management is becoming an increasingly important part of horticulture. Factors that cause the loss of pollinators include pesticide misuse, unprofitability of beekeeping for honey, rapid transfer of pests and diseases to new areas of the globe, urban/suburban development, changing crop patterns, clearcut logging (particularly when mixed forests are replaced by monoculture pine), clearing of hedgerows and other wild areas, loss of nectar corridors for migratory pollinators, and human paranoia of stinging insects.In 1989, following Hurricane Hugo, massive aerial applications for mosquitoes were done in South Carolina. The following year, watermelon growers who did not place beehives in the fields, observed the fruit begin to develop, then abort, or develop into small deformed fruit.

There were entire fields that never yielded a single usable melon. Some growers went out of business; others began to seriously manage pollination. Since beekeepers were also heavily damaged by the mosquito spraying, the supply of bees for pollination was critically short for several years.

IMPORTANCE OF POLLINATION MANAGEMENT

The increasing size of fields and orchards increase the importance of pollination management. Monoculture can cause a brief period when pollinators have more food resources than they can use, while other periods of the year can bring starvation or pesticide contamination of food sources. Most pollinator species rely on a steady nectar source and pollen source throughout the growing season to build up their numbers.

Crops that traditionally have had managed pollination include apple, almonds, pears, some plum and cherry varieties, blueberries, cranberries, cucumbers, cantaloupe, watermelon, alfalfa seeds, onion seeds, and many others.

Some crops that have traditionally depended entirely on chance pollination by wild pollinators need pollination management nowadays to make a profitable crop. Some crops, especially when planted in a monoculture situation, require a very high level of pollinators to produce economically viable crops.

This may be because of lack of attractiveness of the blossoms, or from trying to pollinate with an alternative when the native pollinator is extinct or rare. These include crops such as alfalfa, cranberries, and kiwifruit. This

technique is known as saturation pollination. In many such cases, various native bees are vastly more efficient at pollination but the inefficiency of the honey bees is compensated for by using large numbers of hives, the total number of foragers thereby far exceeding the local abundance of native pollinators.

In a very few cases, it has been possible to develop commercially viable pollination techniques that use the more efficient pollinators, rather than continued reliance on honey bees, as in the management of the alfalfa leafcutter bee.

Table. Number of hives needed per unit area of crop pollination

Common name	Number of hives per acre	Number of hives per hectare	Number of bee visits per square meter/minute
Alfalfa	1, (3–5)	2.5, (4.9–12)	
Almonds	2–3	4.9–7.4	
Apples (normal size)	1	2.5	
Apples (semi dwarf)	2	4.9	
Apples (dwarf)	3	7.4	
Apricots	1	2.5	
Blueberries	3–4	7.4–9.9	2.5
Borage	0.6–1.0	1.5–2.5	
Buckwheat	0.5–1	1.2–2.5	
Canola	1	2.5	
Canola (hybrid)	2.0–2.5	4.9–6.2	
Cantaloupes	2–4, (average 2.4)	4.9–9.9, (average 5.9)	
Clovers	1–2	2.5–4.9	
Cranberries	3	7.4	
Cucumbers	1–2, (average 2.1)	2.5–4.9, (average 5.2)	
Ginseng	1	2.5	
Muskmelon	1–3	2.5–7.4	
Nectarines	1	2.5	
Peaches	1	2.5	
Pears	1	2.5	
Plums	1	2.5	
Pumpkins	1	2.5	
Raspberries	0.7–1.3	1.7–3.2	
Squash	1–3	2.5–7.4	
Strawberries	1–3.5	2.5–8.6	
Sunflower	1	2.5	
Trefoil	0.6–1.5	1.5–3.7	
Watermelon	1–3, (average 1.3)	2.5–4.9, (average 3.2)	
Zucchini	1	2.5	

It is estimated that about one hive per acre will sufficiently pollinate watermelons. In the 1950s when the woods were full of wild bee trees, and beehives were normally kept on most South Carolina farms, a farmer who grew ten acres (4 ha) of watermelons would be a large grower and probably had all the pollination needed.

But today's grower may grow 200 acres and, if lucky, there might be one bee tree left within range. The only option in the current economy is to bring beehives to the field during blossom time.

TYPES OF POLLINATORS

Organisms that are currently being used as pollinators in managed pollination are honey bees, bumblebees, alfalfa leafcutter bees, and orchard mason bees. Other species are expected to be added to this list as this field develops. Humans also can be pollinators, as the gardener who hand pollinates her squash blossoms, or the Middle Eastern farmer, who climbs his date palms to pollinate them.

The Cooperative extension service recommends one honey bee hive per acre for standard watermelon varieties to meet this crop's pollination needs. In the past, when fields were small, pollination was accomplished by a mix of bees kept on farms, bumblebees, carpenter bees, feral honey bees in hollow trees and other insects. Today, with melons planted in large tracts, the grower may no longer have hives on the farm; he may have poisoned many of the pollinators by spraying blooming cotton; he may have logged off the woods, removing hollow trees that provided homes for bees, and pushed out the hedgerows that were home for solitary native bees and other pollinating insects.

PLANNING FOR IMPROVED POLLINATION

Before pollination needs were understood, orchardists often planted entire blocks of apples of a single variety. Because apples are self sterile, and different members of a single variety are genetic clones, this is not a good idea. Growers now supply pollenizers, by planting crab apples interspersed in the rows, or by grafting crab apple limbs on some trees. Pollenizers can also be supplied by putting drum bouquets of crab apples or a compatible apple variety in the orchard blocks.

The field of pollination management cannot be placed wholly within any other field, because it bridges several fields. It draws from horticulture, apiculture, zoology, ecology, and botany.

SHEET MULCHING

Agriculture with mulch in the tropics promotes plant health and vigor. Mulching improves nutrient and water retention in the soil, encourages favorable soil microbial activity and worms, and suppresses weed growth. When properly executed, mulching can significantly improve the well-being of plants and reduce maintenance as compared to bare soil culture. Mulched plants have better Vigour and, consequently have improved resistance to pests and diseases.

"Mulch" is a layer of decaying organic matter on the ground. Mulch occurs naturally in all forests; it is a nutrient rich, moisture absorbent bed of decaying forest leaves, twigs and branches, teeming with fungal, microbial and insect life. Natural mulch serves as a "nutrient bank," storing the nutrients contained in organic matter and slowly making these nutrients available to plants. All forms of plant life from the ground layer to shrubs and trees thrive, grow, shed organic matter, die and decay, in a complicated cycle of nutrients.

Mulch forms a necessary link in nutrient cycling vital for our soils. When mulch is absent for whatever reason, the living soil is robbed of its natural nutrient stores, becomes leached and often desiccates. Natural environments without a litter layer are usually deserts. Non-desert plants grown in bare soil require constant fertilization, nutrient amendment and water, not to mention the work required to keep the soil bare.

Sheet mulching as described here is a suggested method for controlling weeds and improving soil and plant health with mulch. The process mimics the litter layer of a forest floor.

BASIC TECHNIQUES OF SHEET MULCHING

Once you get the hang of it, sheet mulching can be used anywhere plants are grown in the ground. Sheet mulching may be used either in establishing a new garden or tree planting, or to enrich existing plantings. In both cases, mulch is applied to bare soil or on top of weeds. New plantings are planted through the mulch, and a small area is left open to accommodate established plants and trees.

The benefits of mulching justify putting the energy into doing the job right, using ample materials. Collect all of the materials and complete the mulching process in a day. A reduction in maintenance and increase in plant Vigour will reward the initial effort.

Sheet mulch is put down in layers to mimic natural forest mulch: well decayed compost, weed barrier, partly decayed compost and raw organic matter.

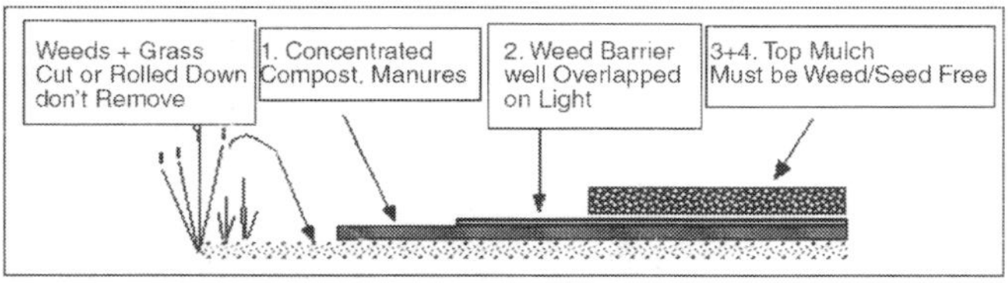

HOW TO SHEET MULCH

Step 1: The Concentrated Compost Layer

To prepare the site, knock down tall weeds and woody plants with a brush cutter, scythe, or simply by trampling the area. Then proceed to lay down the sheet mulch.

Whether you are mulching bare soil or weeds, the first step is to "jump start" microbial activity by adding enriched compost, poultry or stock manure, worm castings or the like at the rate of about 50 lbs/100 square feet. This high nitrogen matter stimulates soil life and gets things going. If the soil is acid, which it likely is if the area has been disturbed recently and treated with conventional fertilizers, add a layer of lime or crushed coral. A soil analysis will indicate the need for adjustment of pH or mineral amendments. This is the appropriate time to add the recommended doses of amendments such as rock phosphate and K mag. Soak the area well with water when the amendments are dispersed.

Step 2: The Weed Barrier

Most cultivated areas today Harbour untold numbers of weed seeds. There are also weed seeds carried around by wind, animals and people. Soil borne seeds are lying dormant and waiting for the right conditions to sprout. Simply pulling or killing growing weeds will not erase the weed problem: more seeds will sprout almost as soon as the soil is exposed to moisture and light. Therefore the next step in mulching is to put down an organic weed barrier. This barrier prevents the germination and eventual emergence of weeds through your mulch.

Underneath this weed barrier grasses and weeds die and quickly become food for earthworms. From now on, the worms turn and aerate the soil, as they do naturally when in the right environment.

Of the four sheet mulch layers, the weed barrier has no natural counterpart on the forest floor. In the forest, weeds do not sprout because there is "no room for them," which simply means a lack of space above and below the ground, and a lack of light. By planting an area properly, there will eventually be no room for weeds. The weed barrier is needed only for establishment of the mulch, and disappears with time. If your area is planted appropriately, weeds will not emerge after the decomposition of the weed barrier.

Materials for the weed barrier that work well are: 4-6 sheets of newspaper, cardboard, burlap bags, old rugs of natural Fibre, worn-out jeans, gypsum board, or whatever you can find around. Banana, ape and ti leaves also work if laid down in several layers. Overlap the pieces of the material so as to completely cover the ground without any breaks, except where there are plants you want to save. Around these leave a generous opening for air circulation around the root crown. Care in laying down the weed barrier will save you the headache of emerging weeds later on.

Step 3: The Compost Layer

This layer is on top of the weed barrier - it must be weed seed free. Well conditioned compost, grass clippings, seaweed and leaves are ideal materials to spread over the weed barrier. Any weed-free material mixture at the right moisture level for a good compost will do. This should form a fairly dense layer about 3 inches thick.

Step 4: The Top Layer

The top dressing mimics the newly fallen organic matter of the forest. It also must be weed-free. Good materials for this include leaves, twigs and small branches, fern or palm fronds, straw, coffee chaff, macadamia nut shells, wood chips, sawdust, bark, etc.. The top layer will slowly decompose into lower layers, and therefore must be replaced periodically; it represents reserves of compost. This layer should be about 3-5 inches deep. Many materials suitable for the top layer often have a pleasant cosmetic appearance. What luck! For this reason, there should be no hesitation in using sheet mulch in all cultivation from landscaping to gardening to permanent orchard crops. In fact, as you use mulch, bare soil will begin to seem ugly and undesirable. When the soil is amended and sheet mulch applied properly, there will never be a need to turn the soil. Earthworms do the tilling. The only task left will be to keep the soil covered by replenishing the mulch.

Warning: Feral pigs love good, moist soil, and will grub in sheet mulch if they have access to it. Do not use sheet mulch if pigs have access to the area; they will be attracted to it and will destroy both your work and your plantings.

Mulching Around Trees

- Plant tree.
- Amend soil around tree in a wide ring shape from a few centimeters from trunk out to 1 meter with a light layer of nitrogen fertilizer, such as chicken manure, and other amendments if necessary. Rake or water in.
- Spread a layer of permeable weed barrier around the tree in a ring shape, leaving about 15 cm diameter around the trunk of the tree for it to "breathe." Make certain there are no gaps in the ring shape through which weeds can emerge. Water the weed barrier layer thoroughly before the next step.
- Spread compost and/or mulch about 15 cm thick over the weed barrier, again making sure it is several centimeters away from the trunk of the plant.

The Ongoing Process

To make mulching as efficient and easy as possible, use mulch materials which are readily available. With good planning, mulching of gardens and orchards can become a regular part of maintenance-just mulch with handy materials such as grass clippings, plant prunings animal bedding, etc.. Eventually, other tasks such as watering, fertilization and weeding will be reduced. The overall maintenance burden in mulched conditions, when properly executed, is far less than in conventional systems.

Once a plant is properly mulched, its own leaf drop will constantly add to that mulch. But is natural leaf drop enough to maintain the mulch? The answer to this depends on the plant species and also how the plant is growing in relation to other plants. Certain trees produce tremendous amounts of leaf matter which decomposes rather slowly; examples are: avocado, macadamia, lychee, as well as many others. These trees can be expected to generate sufficient mulch for themselves once vigorous growth is attained. Unfortunately, under most conditions many trees do not create enough long lasting mulch for maintenance of their needs. To explain this apparent deficiency, look once again at the forest. Here, plants are "stacked" in the vertical direction in ground-level, middle, and tall vegetation. This means that the ground under each plant receives organic matter from several plants. There are many ways to produce sufficient mulch at your site. Grass clippings represent nutrient rich mulch material. Deep rooted, vigorous growing plants that readily come back from hard pruning or coppicing will also work. There are several nitrogen fixing trees which produce copious amounts of green matter. Each should be evaluated for the specific site before planting. Other plants that work well are kukui, hau, desmodium,, various bunch grasses, lemon grass, comfrey, etc.. Also, many water plants such as water hyacinth are good mulch materials. Since plants that produce heavy amounts of organic matter are by their nature nearly irrepressible, extreme caution should be taken not to let these plants escape your management and become weedy.Sheet mulching should not be confused with composting, artificial weed barriers, or green manuring. Sheet mulching as described here is quite different from these in that it seeks to recreate the organic mulch layer of the forest with a minimum of effort from people. Properly planned, a backyard or orchard system will produce its own raw mulch in sufficient amounts and people are involved only in putting this material back onto the ground where it belongs.

SQUARE FOOT GARDENING

Square Foot Gardening is the practice of planning small but intensively planted gardens. The phrase "square foot gardening" was popularized by Mel Bartholomew. The practice combines concepts from other organic gardening methods, including a strong focus on compost, closely planted raised beds and biointensive attention to a small, clearly defined area. Proponents claim that the method is particularly well-suited for areas with poor soil, beginning gardeners or as adaptive recreation for those with disabilities.

BASIC METHOD

The original square-foot-gardening method used an open-bottomed box to contain a finite amount of soil, which was divided with a grid into sections. To encourage variety of different crops over time, each square would be planted with a different kind of plant, the number of plants per square depending on an individual plant's size. A single tomato plant might take a full square, as might herbs such as oregano, basil or mint, while most strawberry plants could be planted four per square, with up to sixteen radishes per square. Tall or climbing plants such as maize or pole beans might be planted in a northern row so as not to shade other plants, and supported with lattice or netting.

The logic behind using smaller beds is that they are easily adapted, and the gardener can easily reach the entire area, without stepping on and compacting the soil. In the second edition, Bartholomew suggests using a "weed barrier" beneath the box, and filling it completely with "Mel's mix," a combination by volume of one third of decayed Sphagnum "peat moss", one-third expanded vermiculite and one-third blended compost. For accessibility, raised boxes may have bottoms to sit like tables at a convenient height, with approximately 6" of manufactured soil per square foot.

RATIONALE

In this method, the garden space is divided into beds that are easily accessed from every side. A 4' x 4', 16 sq ft or 120 cm x 120 cm, 1.4 m^2 garden is recommended for the first garden, and a path wide enough to comfortably work from should be made on each side of the bed, if possible, or if the bed must be accessed by reaching across it, a narrower one should be used so that no discomfort results from tending the garden. Each of the beds is divided into approximately one square foot units and marked out with sticks, twine, or sturdy slats to ensure that the square foot units remain visible as the garden matures.

Different seeds are planted in each square, to ensure a rational amount of each type of crop is grown, and to conserve seeds instead of overplanting, crowding and thinning plants. Common spacing is one plant per square for larger plants, four plants per square for medium large plants like lettuce, nine plants per square for medium-small plants like spinach, and sixteen per square for small plants such as onions and carrots. Plants that normally take up yards of space as runners, such as squash or cucumbers, are grown vertically on sturdy frames that are hung with netting or string to support the developing crops. Ones that grow deep underground, such as potatoes or carrots, are grown in a square foot section that has foot tall sides and a planting surface above the ground, so that a foot or more of framed soil depth is provided above the garden surface rather than below it.

The beds are weeded and watered from the pathways, so the garden soil is never stepped on or compacted. Because a new soil mixture is used to create the garden, and a few handfuls of compost are added with each harvest to maintain soil fertility over time, the state of the site's underlying soil is irrelevant. This gardening method has been employed successfully in every region, including in deserts, on high arid mountain plateaus, in cramped urban locations, and in areas with polluted or high salinity soils. It is equally useful for growing flowers, vegetables, herbs and some fruits in containers, raised beds, on tabletops or at ground level, in only 4 to 6 inches of soil. A few seeds per square foot, the ability to make compost, to water by hand, and to set up the initial garden in a sunny position or where a container, table or platform garden may be moved on wheels to receive light is all that is needed to set up a square foot garden.

BENEFITS OF SQUARE FOOT GARDENING

- *Much less work*: Conventional gardening requires heavy tools to loosen the soil, whereas in this method, the soil is never compacted and it remains loose and loamy. Weeding takes only seconds to minutes, due to the light soil, raised beds, and easily accessed plants. Harvests per foot of garden are increased due to the rich soil mixture, well-spaced plants, and lack of weeds produced when following Mel Bartholomew's method.

- *Water Savings*: The soil mixture that is advised has water-holding capacities, so that the garden needs water less frequently, and in much smaller quantities than when using other gardening methods. Water is also spared by hand-watering directly at the plant roots, so that there is very little waste and tender young plants and seedlings are preserved.
- *Very little weeding*: One benefit of this close planting is that the vegetables form a living mulch, and shade out many weed seeds before they have a chance to germinate.
- *Pesticide / Herbicide Free*: Natural insect repellent methods like companion planting become very efficient in a close space and thus, pesticides are not necessary. The large variety of crops in a small space also prevents plant diseases from spreading easily.
- *Accessibility*: A plywood bottom can be attached to the bottom of a box, which can then be placed on a tabletop or raised platform for those who wish to garden without bending or squatting, or to make gardening easy for wheelchair, cane or walker users.

SUCCESSION PLANTING

Succession planting is done so garden space is put to its most efficient use. With this method, there is never an unplanted area in the vegetable plot. As soon as one crop is harvested, the space is immediately replanted with a different crop. This method of succession planting uses groupings of cool-season and warm-season crops. In the simplest plan, succession planting involves planning for spring, summer and fall crops.

Another method is to make several plantings of a vegetable. Bush beans can be planted every 2 weeks from mid-May to the beginning of August. A third type of succession planting involves a single planting of several different cultivars with varying maturity dates. This method works well for cabbage and corn. These methods allow for a continuous harvest over a longer period.

Vegetable harvest can be extended into the fall by growing the faster growing, cool-season crops. The quality of many of the cool-season crops is exceptional when they mature in the fall. Plant so there is adequate time for the crop to mature.

When planting seeds for a fall crop, you must water the seeds more often than in the spring because the summer soil is usually warm and dry. Provide some shade to help retain moisture. Cover the seedbed with cardboard or newspaper. Check the seedbed daily and remove the covering when the first seed sprouts. Keep the bed consistently watered until all the seedlings are established. Pre-sprouting also ensures good germination of cool-season crops in the hot, dry soil of August.

WEED CONTROL

Weed control is the botanical component of pest control, using physical and chemical methods to stop weeds from reaching a mature stage of growth when they could be harmful to domesticated plants and livestock. In order to reduce weed growth, many "weed control" strategies have been developed in order to contain the growth and spread of weeds.

The most basic is ploughing which cuts the roots of annual weeds. Today, chemical weed killers known as herbicides are widely used.

THE EFFECTS OF WEEDS ON OTHER PLANTS

Weeds can compete with productive crops or pasture, or convert productive land into unusable scrub. Weed are also often poisonous, distasteful, produce burrs, thorns or other damaging body parts or otherwise interfere with the use and management of desirable plants by contaminating harvests or excluding livestock.

Weeds tend to thrive at the expense of the more refined edible or ornamental crops. They provide competition for space, nutrients, water and light, although how seriously they will affect a crop depends on a number of factors. Some crops have greater resistance than others- smaller, slower growing seedlings are more likely to be overwhelmed than those that are larger and more vigorous. Onions are one of the crops most susceptible to competition, for they are slow to germinate and produce slender, upright stems. Quick growing, broad leafed weeds therefore have a distinct advantage, and if not removed, the crop is likely to be lost. Broad beans however produce large seedlings, and will suffer far less profound effects of weed competition other than during periods of water shortage at the crucial time when the pods are filling out. Transplanted crops raised in sterile seed or potting compost will have a head start over germinating weed seeds.

Weeds also differ in their competitive abilities, and can vary according to conditions and the time of year. Tall growing vigorous weeds such as fat hen can have the most pronounced effects on adjacent crops, although seedlings of fat hen that appear in late summer will only produce small plants. Chickweed, a low growing plant, can happily co-exist with a tall crop during the summer, but plants that have overwintered will grow rapidly in early spring and may swamp crops such as onions or spring greens.The presence of weeds does not necessarily mean that they are competing with a crop, especially during the early stages of growth when each plant can find the resources it requires without interfering with the others. However, as the seedlings' size increases, their root systems will spread as they each begin to require greater amounts of water and nutrients. Estimates suggest that weed and crop can co-exist harmoniously for around three weeks, therefore it is important that weeds are removed early on in order to prevent competition occurring. Weed competition can have quite dramatic effects on crop growth.

Harold A Roberts cites research carried out with onions wherein "Weeds were carefully removed from separate plots at different times during the growth of the crop and the plots were then kept clean.

It was found that after competition had started, the final yield of bulbs was being reduced at a rate equivalent to almost 4% per day. So that by delaying weeding for another fortnight, the yield was cut to less than half that produced on ground kept clean all the time." He goes on to record that "by early June, the weight of weeds per unit area was twenty times that of the crop, and the weeds had already taken from the soil about half of the nitrogen and a third of the potash which had been applied".

Perennial weeds with bulbils, such as lesser celandine and oxalis, or with persistent underground stems such as couch grass or creeping buttercup are able to store reserves of food, and are thus able to grow faster and with more vigour than their annual counterparts. There is also evidence that the roots of some perennials such as couch grass exude allelopathic chemicals which inhibit the growth of other nearby plants.

Weeds can also host pests and diseases that can spread to cultivated crops. Charlock and Shepherd's purse may carry clubroot, eelworm can be harboured by chickweed, fat hen and shepherd's purse, while the cucumber mosaic virus, which can devastate the cucurbit family, is carried by a range of different weeds including chickweed and groundsel.However, at times the role of weeds in this respect can be over-rated.

As far as insect pests are concerned, often the species that live on weeds are not the same as those that attack vegetable crops; "Tests with the common cruciferous weeds such as shepherds purse have shown that they do not act as hosts for the larvae of the cabbage root fly. One exception was found to be the wild radish, but this is not usually a weed of established vegetable gardens". However pests such as cutworms may first attack weeds then move on to cultivated crops.

While charlock, a common weed in southeastern USA, may be considered a weed by row crop growers, it is highly valued by beekeepers, who seek out places where it blooms all winter, thus providing pollen for honeybees and other pollinators. Its bloom is resistant to all but a very hard freeze, and even that will only kill it back briefly. By feeding an array of pollinators during a seasonal dearth, it can redound to the farmer's advantage. Many weeds are likewise highly beneficial to pollinators.

METHODS

In domestic gardens, methods of weed control include covering an area of ground with several layers of wet newspaper or one black plastic sheet for several weeks. In the case of using wet newspaper, the multiple layers prevent light from reaching all plants beneath, which kills them. Saturating the newspaper with water daily speeds the decomposition of the dead plants. Any weed seeds that start to sprout because of the water will also be deprived of sunlight, be killed, and decompose. After several weeks, all germinating weed seeds present in the ground should be dead. Then the newspaper can be removed and the ground can be planted. The decomposed plants will help fertilise the plants or seeds planted later.

In the case of using the black plastic sheet, the greenhouse effect is used to kill the plants beneath the sheet. A 5-10 cm layer of wood chip mulch on the ground will also prevent most weeds from sprouting. Also, gravel can be spread over the ground as an inorganic mulch. In agriculture, irrigation is sometimes used as a weed control measure such as in the case of paddy fields. Many people find that although the black plastic sheeting is extremely effective at preventing the weeds in areas where it covers, in actual use it is difficult to achieve full coverage.

Knowing how weeds reproduce, spread and survive adverse conditions can help in developing effective control and management strategies. Weeds have a range of techniques that enable them to thrive;

Annual and biennial weeds such as chickweed, annual meadow grass, shepherd's purse, groundsel, fat hen, cleaver, speedwell and hairy bittercress propagate themselves by seeding. Many produce huge numbers of seed several times a season, some all year round. Groundsel can produce 1000 seed, and can continue right through a mild winter, whilst Scentless Mayweed produces over 30,000 seeds per plant. Not all of these will germinate at once, but over several seasons, lying dormant in the soil sometimes for years until exposed to light. Poppy seed can survive 80-100 years, dock 50 or more. There can be many thousands of seeds in a square foot or square metre of ground, thus and soil disturbance will produce a flush of fresh weed seedlings.

"Stale Seed Bed" Technique

One technique employed by growers is the 'stale seed bed', which involves cultivating the soil, then leaving it for a week or so.

When the initial flush of weeds has germinated, the grower will lightly hoe off before the desired crop is planted. However, even a freshly cleared bed will be susceptible to airborne seed from elsewhere, as well as seed brought in by passing animals which can carry them on their fur, or from freshly imported manure. The organic solution to the problem of spreading annual weeds lies in regular, properly timed weeding, preferably just before flowering.

This technique is also quite often used by farmers who let weeds germinate then return the soil before crop sowing.Perennial weeds also propagate by seeding; the airborne seed of the dandelion and the rose-bay willow herb are parachuted far and wide. But they also have an additional range of vegetative means of spreading that gives them their pernicious reputation. Dandelion and dock put down deep tap roots, which, although they do not spread underground, are able to regrow from any remaining piece left in the ground. Removal of the complete tap root is the only sure remedy.

The most persistent of the perennials are those that spread by underground creeping rhizomes that can regrow from the tiniest fragment. These include couch grass, bindweed, ground elder, nettles, rosebay willow herb, Japanese knotweed, horsetail and bracken, as well as creeping thistle, whose tap roots can put out lateral roots. Other perennials put out runners that spread along the soil surface. As they creep along they set down roots, enabling them to colonise bare ground with great rapidity. These include creeping buttercup and ground ivy. Yet another group of perennials propagate by stolons- stems that arch back into the ground to reroot. Most familiar of these is the bramble.

All of the weeds can be very difficult to eradicate- thick black plastic mulches can be effective to a degree, although will probably need to be left in place for at least two seasons. In addition, hoeing off weed leaves and stems as soon as they appear can eventually weaken and kill the plants, although this will require persistence in the case of plants such as bindweed. Nettle infestations can be tackled by cutting back at least three times a year, repeated over a three year period. Bramble can be dealt with in a similar way. Some plants are said to produce root exudates that suppress herbaceous weeds. Tagetes minuata is claimed to be effective against couch and ground elder, whilst a border of comfrey is also said to act as a barrier against the invasion of some weeds including couch.

Use of Herbicides

The methods of weed control avoid using chemicals. They are often used by farmers. However, these methods may damage a fragile soil by restructuring it, hence are not always used. They are those preferred by the organic gardener or organic farmer.However weed control can also be achieved by the use of herbicides. Selective herbicides kill certain targets while leaving the desired crop relatively unharmed. Some of these act by interfering with the growth of the weed and are often based on plant hormones.

Herbicides are generally classified as follows:

- Contact herbicides destroy only that plant tissue in contact with the chemical spray. Generally, these are the fastest acting herbicides. They are ineffective on perennial plants that are able to re-grow from roots or tubers.

- Systemic herbicides are foliar-applied and are translocated through the plant and destroy a greater amount of the plant tissue. Modern herbicides such as glyphosate are designed to leave no harmful residue in the soil.
- Soil-borne herbicides are applied to the soil and are taken up by the roots of the target plant.
- Pre-emergent herbicides are applied to the soil and prevent germination or early growth of weed seeds.

In agriculture large scale and systematic weeding is usually required, often by machines, such as liquid herbicide sprayers, or even by helicopter, to eliminate the massive amount of weeds present on farming lands.

However there are a number of techniques that the organic farmer can employ such as mulching and carefully timed cutting of weeds before they are able to set seed.

Organic Methods

Typically a combination of methods are used in organic situations:

- *Drip irrigation*: Rubber hoses and other methods are used to bring water directly to the roots of the desired plants. This limits weed access to water.
- *Manually pulling weeds*: Labourers are used to pull weeds at various points in the growing process.
- *Mechanically tilling around plants*: Tractors are used to carefully till weeds around the crop plants at various points in the growing process. Besides tilling, other mechanical weed control methods also exist
- *Ploughing*: Ploughing includes tilling of soil, intercultural ploughing and summer ploughing. Ploughing through tilling of soil uproots the weeds which causes them to die. In summer ploughing is done during deep summers. Summer ploughing also helps in killing pests.
- *Crop rotation*: Rotating crops with ones that kill weeds by choking them out, such as hemp, Mucuna pruriens, and other crops, can be a very effective method of weed control. It is a way to avoid the use of herbicides, and to gain the benefits of crop rotation.
- *Weed mat*: A weed mat is an artificial mulch, fibrous cloth material, bark or newspaper laid on top of the soil preventing weeds from growing to the surface.

WOOD

Besides those kinds of weeds which are of an herbaceous nature, there are others which are woody, and grow to a very considerable size; such as broom, furze and Lantanas. The first may be destroyed by frequent ploughing and harrowing, in the same manner as other perennial weeds are. Another method of destroying broom is by pasturing the field where it grows with grazers.The best method of extirpating furze is to set fire to it in frosty weather, for frost has the effect of withering and making them burn readily. The stumps must then be cut over with a hatchet, and when the ground is well softened by rain it may be ploughed up, and the roots taken out by a harrow adapted to that purpose. If the field is soon laid down to grass, they will again spring up; in this case, pasturing with grazers is an effectual remedy. The thorn, or bramble, can only be extirpated by ploughing up the ground and collecting the roots.

In June, weeds are in their most succulent state, and in this condition, after they have lain a few hours to wither, cattle will eat almost every species. There is scarcely a hedge, border, or a nook, but what at that season is valuable.

UK LEGISLATION CONCERNING WEEDS

The Weeds Act, 1959 is described as "Preventing the spread of harmful or injurious weeds", and is mainly relevant to farmers and other rural settings rather than the allotment or garden scale grower. There are five 'injurious' weeds covered by the provisions of the Weeds Act.

These are:

- Spear thistle
- Creeping or field thistle
- Curled Dock
- Broad leaved dock
- Common ragwort.

The Department for Environment, Food and Rural Affairs provides guidance for the treatment removal of these weeds from infested land. Much of this is oriented towards the use of herbicides, the majority of which may not be acceptable to the organic producer but in most cases there are manual techniques that can be used such as digging out the roots, mulching out or carefully timed cutting before seeds are able to spread.

Primary responsibility for weeds control rests with the occupier of the land on which the weeds are growing, therefore it is important to be alert to potential weed problems and to take prompt action. However, it should be remembered that most common farmland weeds are not "injurious" within the meaning of the Weeds Act, and many such plant species have conservation and environmental value. When dealing with complaints under the Weeds Act, DEFRA has a duty in law to try and achieve a reasonable balance between different interests. These include agriculture, countryside conservation and the public in general. Constructive discussion about problems caused by weeds can often result in effective solutions and avoid the need for DEFRA to take official action. In addition to those weeds covered by the 1959 act, under section 14 of the Wildlife and Countryside Act 1981, it can be an offence to plant or grow certain specified plants in the wild including Giant Hogweed and Japanese

Knotweed. Problems involving these plants can be referred to the local authority for the area where those weeds are growing as some local authorities have bye-laws controlling these plants. There is no statutory requirement for landowners to remove these plants from their property.

STALE SEED BED

A false or stale seed bed is a useful organic weed control technique which involves creating a seedbed some weeks before seed is due to be sown. Preparation of such seedbed makes sure that any weed seeds that have been disturbed and brought to the soil surface during cultivation will thus have a chance to germinate, and can then be hoed off before sowing of the actual crop is carried out.

XERISCAPING

Xeriscaping and xerogardening refers to landscaping and gardening in ways that reduce or eliminate the need for supplemental irrigation. It is promoted in areas that do not have easily accessible supplies of fresh water, and is gaining acceptance in other areas as climate patterns shift.The word xeriscaping is a portmanteau of xeros and landscape.Plants whose natural requirements are appropriate to the local climate are emphasized, and care is taken to avoid losing water to evaporation and run-off. While many Denver Water employees helped coin the term xeriscape, Xeriscape and the xeriscape logo are not registered trademarks of Denver Water, the water department of Denver, Colorado. They were created by the Front Range Xeriscape Task Force of Denver Department in 1978. The specific plants used in xeriscaping depend upon the climate. Some common plants used in Western xeriscaping are agave, cactus, lavender, juniper, sedum and thyme.In some areas, terms such as water-conserving landscapes, drought-tolerant landscaping, zeroscaping, and smart scaping are used instead.

ADVANTAGES

- Lower water consumption
- More water available for other uses and other people
- Less time and work needed for maintenance, making gardening simpler and less stressful
- Little or no lawnmowing
- Xeriscape plants along with proper bed design tends to take full advantage of rainfall
- When water restrictions are implemented, xeriscape plants will tend to survive, while more traditional plants may be unable to adapt.

GREYWATER

Greywater is wastewater generated from domestic activities such as laundry, dishwashing, and bathing which can be recycled on-site for uses such as landscape irrigation and constructed wetlands. Greywater differs from water from the toilets which is designated sewage or blackwater to indicate it contains human waste.

Greywater composes 50–80% of residential wastewater generated from all of the house's sanitation equipment.

Greywater gets its name from its cloudy appearance and from its status as being neither fresh, nor polluted. Wastewater containing significant food residues or high concentrations of toxic chemicals from household cleaners, etc., may be considered "dark grey".

CHALLENGES

In recent years, concerns over dwindling reserves of groundwater and overloaded or costly sewage treatment plants have generated much interest in the reuse or recycling of greywater, both domestically and for use in commercial irrigation.

A lack of knowledge and experience with modern greywater systems has led to some concern over potential health and environmental risks. If properly designed a greywater system will lower sewage costs, at the same time reducing water usage for irrigation and water quality can be increased to better than the government supplied initially. Less water enters the city sewage systems saving the government from building new treatment plants. Some jurisdictions demand highly intensive treatment systems for legal reuse of greywater whereas some encourage greywater use. Oregon, on June 12, 2009, passed a law which legalizes and encourages the use of greywater. Despite many obstacles, greywater is often reused for irrigation legally and illegally.

Anyone, including those in droughtzones or areas hit by hose pipe bans can harvest greywater by manual bucketing shower or sink water and using them to flush toilets. In the third world, reuse of greywater is often unregulated and is common. At present, the recycling of greywater is poorly understood by many compared with elimination.

ELIMINATION OF GREYWATER

Domestic wastewater is usually combined at the sewer, so that grey- and blackwaters are removed together using a shared sewerage system in a process called elimination.

Sewage water can then be treated to limit pollution and health risks, before being returned to the environment at large. Most greywater ends up as effluent in rivers and oceans in this way.

There are other alternatives to eliminating greywater that allow for efficient use; using it to irrigate plants is a common practice. The plants use contaminants of greywater, such as food particles, as nutrients in their growth. However, salt and soap residues can be toxic to microbial and plant life alike, but can be absorbed and degraded through constructed wetlands and aquatic plants such as sedges, rushes, and grasses.

RECYCLING

Most greywater is easier to treat and recycle than blackwater, because of lower levels of contaminants. While all greywater contains micro-organisms, the health hazards associated with greywater from a multiple-dwelling source should be considered differently from that of a single dwelling greywater source. One may recycle greywater at home by putting a bucket in the shower or sink and using the water to flush the toilet.

If collected using a separate plumbing system from blackwater, domestic greywater can be recycled directly within the home, garden or company and used either immediately or processed and stored. Recycled greywater of this kind is never safe to drink, but a number of stages of filtration and microbial digestion can be used to provide water for washing or flushing toilets. Some greywater may be applied directly from the sink to the garden or container field, receiving further treatment from soil life and plant roots. Given that greywater may contain nutrients, pathogens, and is often discharged warm, it is very important not to store it before use in irrigation purposes, unless it is properly treated first.

Systems

At present, several water recycling systems exist which can be used to:

- Recycle the water without purifying it
- Recycle the water while purifying or decontaminating it

Water recycling without purification is used in certain agricultural companies and dwellings for applications where potable water is not required. It may also be used in dwellings when the greywater is already fairly clean to begin with and/or has not been polluted with non-degradable chemicals such as non-natural soaps.

Water purification/decontamination systems then again are used for applications where potable water is required.

Water Recycling Systems Without Purification

Water Diversion Systems

The simplest greywater system is to simply divert the water directly to the garden. Regulations change by country and region, but common guidelines for safe usage include not storing the greywater for more than 24 hours, ensuring it cannot pool or run off, and depositing it with subsurface irrigation. Greywater diversion systems can be both designed-in to new homes, or retrofitted to many existing homes. When systems are fully designed, manufactured and installed to relevant standards such as the Australian Watermark standards. Water diversion systems tend to be highly efficient, effective and safe for simple applications where potable water is not required.

Diversion systems can be as basic as running the outlet hose from a washing machine out a window to the garden, or can be designed as a permanent part of the home plumbing. Fully engineered systems incorporate a sump pump and surge tanks and deliver the water through sub-surface irrigation.Greywater from the shower or bath is generally good quality water for the garden. The soap levels at the dilutions typical are actually good for the garden as they are a wetting agent. When laundry greywater is diverted to the garden then the laundry products must be chosen carefully to ensure phosphate and salt levels are low, and the pH balance is neutral.

Basic guidelines are also available from system suppliers. It is essential that greywater is diverted to sewer when garden-unfriendly products are being used.

Water Recycling with Purification

For filtering the water to become potable, there are numerous systems based on soft processes.

These include natural biological principles such as:

- Mechanical systems
- Biological systems and Bio reactors or compact systems as activated sludge systems, biorotors, aerobic and anaerobic biofilters, submerged aerated filters, biorolls

Finally, also used for creating potable water are the "hard", direct processes, such as distillation. There seem to be no commercially available "hard" greywater recovery devices suitable for on-site use in the individual household, even though a number of such technologies exist.

In order to purify the potable water adequately, several of these systems are usually combined to work as a whole. Combination of the systems is done in two to three stages, knowingly a primary and a secondary purification. Sometimes a tertiary purification is also added.

Some municipal sewerage systems recycle a certain amount of grey- and blackwaters using a high standard of treatment, thus providing reclaimed water for irrigation and other uses.

APPLICATION OF RECYCLED GREYWATER

Irrigation

Greywater typically breaks down faster than blackwater and has lower levels of nitrogen and phosphorus. However, all greywater must be assumed to have some blackwater-type components, including pathogens of various sorts. Greywater should be applied below the surface where possible and not sprayed, as there is a danger of inhaling the water as an aerosol.

Long-term research on greywater use on soil has not yet been done and it is possible that there may be negative impacts on soil productivity. If you are concerned about this, avoid using laundry powders; these often contain high levels of salt as a bulking agent.In any greywater system, it is essential to put nothing toxic down the drain—no bleaches, bath salts, artificial dyes, cleansers, and no products containing boron.

It is crucial to use all-natural, biodegradable soaps whose ingredients do not harm plants. Most powdered detergents, and some liquid detergents, are sodium-based, which can inhibit seed-germination and destroy the structure of clay soils.

"Natural" body products often contain substances toxic to humans, including parabens, stearalkonium chloride, phenoxyethanol, polyethelene glycol, and synthetic fragrances.

Indoor Reuse

Recycled greywater from showers and bathtubs can be used for flushing toilets in most European and Australian jurisdictions and in United States jurisdictions that have adopted the International Plumbing Code.

Such a system could provide an estimated 30% reduction in water use for the average household.

The danger of biological contamination is avoided by using:

- A cleaning tank, to eliminate floating and sinking items
- An intelligent control mechanism that flushes the collected water if it has been stored long enough to be hazardous; this completely avoids the problems of filtration and chemical treatment

The Uniform Plumbing Code, adopted in some United States jurisdictions, prohibits greywater use indoors.

Extreme Living Conditions

Greywater use promotes the ability to build in areas unsuitable for conventional treatment, or where conventional treatment is costly. The Mars Desert Research Station uses greywater recycling, and might be used on trips to Mars to reduce water consumption and increase oxygen generation.

Heat Reclamation

Devices are currently available that capture heat from residential and industrial greywater, through a process called drainwater heat recovery, greywater heat recovery, or hot water heat recycling.

Rather than flowing directly into a water heating device, incoming cold water flows first through a heat exchanger where it is pre-warmed by heat from greywater flowing out from such activities as dishwashing, or showering. Typical household devices receiving greywater from a shower can recover up to 60% of the heat that would otherwise go to waste.

Ice Rinks

Lee Valley Ice Centre in Leyton, London, is the first example of the use of a greywater system in an ice arena context, a concept which reduces the otherwise considerable water use associated with such activities.

ECOLOGY

Because greywater use, especially domestically, reduces demand on conventional water supplies and pressure on sewage treatment systems, its use is very beneficial to local waterways. In times of drought, especially in urban areas, greywater use in gardens or toilet systems helps to achieve the goals of ecologically sustainable development.

Benefits

The potential ecological benefits of greywater recycling include:

- Lower fresh water extraction from rivers and aquifers
- Less impact from septic tank and treatment plant infrastructure
- Topsoil nutrification
- Reduced energy use and chemical pollution from treatment
- Groundwater recharge
- Increased plant growth
- Reclamation of nutrients
- Greater quality of surface and ground water when preserved by the natural purification in the top layers of soil than generated water treatment processes

In the U.S. Southwest and the Middle East where available water supplies are limited, especially in view of a rapidly growing population, a strong imperative exists for adoption of alternative water technologies.

Potential Downsides of Greywater Recycling

The Southern Nevada Water Authority has expressed concern that greywater recycling for use in residential washing machines and for watering xeriscapes could encourage greater water use, and reduce the amount of water returned to Lake Mead, a reservoir currently experiencing a drought. Greywater users and advocates dispute these claims.

GOVERNMENTAL REGULATION

Government regulation governing domestic greywater use for landscape irrigation is still a developing area and continues to gain wider support as the actual risks and benefits are considered and put into clearer perspective.

'Greywater' is considered in some jurisdictions to be 'sewage' but in the U.S. states that adopt the International Plumbing Code, it can be used for sub surface irrigation and for toilet flushing, and in states that adopt the Uniform Plumbing Code, it can be used in underground disposal fields that are akin to shallow sewage disposal fields.

Wyoming allows surface and subsurface irrigation and other Non- specific use of greywater under a Department of Environmental Quality policy enacted in March, 2010. California, Utah, New Mexico and some other states allow true subsurface drip irrigation with greywater. Where greywater is still considered sewage, it is bound by the same regulatory procedures enacted to ensure properly engineered septic tank and effluent disposal systems are installed for long system life and to control spread of disease and pollution. In such regulatory jurisdictions, this has commonly meant domestic greywater diversion for landscape irrigation was either simply not permitted or was discouraged by expensive and complex sewage system approval requirements. Wider legitimate community greywater diversion for landscape irrigation has subsequently been handicapped and resulted in greywater reuse continuing to still be widely undertaken by householders outside of and in preference to the legal avenues.

However, with water conservation becoming a necessity in a growing number of jurisdictions, business, political and community pressure has made regulators seriously reconsider the actual risks against actual benefits.

It is now recognized and accepted by an increasing number of regulators that the microbiological risks of greywater reuse at the single dwelling level where inhabitants already had intimate knowledge of that greywater are in reality an insignificant risk, when properly managed without the need for complex, expensive and onerous red tape approval processes. This is reflected in the NSW Government Department of Water and Energy's newly released greywater diversion rules, and the recent passage of greywater legislation in Montana. In the 2009 Legislative Session, the state of Montana passed a bill expanding greywater use into multi-family and commercial buildings. The Department of Environmental Quality has already drafted rules and design guidelines for greywater re-use systems in all these applications. Existing staff would review systems proposed for new subdivisions in conjunction with review of all other wastewater system components.

UNDERUTILIZED CROPS

Neglected and underutilized crops are plant species that are used traditionally for their food, fibre, fodder, oil or medicinal properties, but have yet to be adopted by large-scale agriculturalists. They may have the potential to contribute to food security, nutrition, health, income generation and environmental services. For a variety of reasons, some economic and some cultural, they have been neglected or underutilized. Just three crops - maize, wheat and rice - account for about half of the world's consumption of protein and calories. 95% of the world's food needs are provided for by just 30 species of plants. In stark contrast, an estimated 7,500 species are considered edible in the world today. Underutilized and neglected plants are those that could be - and, in many cases, historically have been - used for food on a larger scale.

Alongside their commercial potential, many of the underutilized crops also provide important environmental services, as they are adapted to marginal soil and climate conditions.

To be considered as an "underutilized crop", a plant must meet three criteria:

1. It must have a proven food or energy value;
2. It must be proven able to be cultivated, either:
   - By having been widely cultivated in the past, or
   - By only being cultivated in a limited geographical area; and,
3. It must be currently cultivated less than other comparable plants.

4

Farming in India

Sustainable development has caught the imagination and action all over the world for more than a decade. Sustainable agriculture is necessary to attain the goal of sustainable development. Food and Agriculture Organization, sustainable agriculture "is the successful management of resources for agriculture to satisfy changing human needs while maintaining or enhancing the quality of environment and conserving natural resources". All definitions of sustainable agriculture lay great emphasis on maintaining an agriculture growth rate, which can meet the demand for food of all living things without draining the basic resources.

Farming is one of the several approaches found to meet the objectives of sustainable agriculture. Many techniques used in farming like inter-cropping, mulching and integration of crops and livestock are not alien to various agriculture systems including the traditional agriculture practiced in old countries like India. However, farming is based on various laws and certification programmes, which prohibit the use of almost all synthetic inputs, and health of the soil is recognised as the central theme of the method.Adverse effects of modern agricultural practices not only on the farm but also on the health of all living things and thus on the environment have been well documented all over the world. Application of technology, particularly in terms of the use of chemical fertilizers and pesticides all around us has persuaded people to think aloud. Their negative effects on the environment are manifested through soil erosion, water shortages, salination, soil contamination, genetic erosion, etc.

Farming is one of the widely used methods, which is thought of as the best alternative to avoid the ill effects of chemical farming. There are several definitions of farming and the one given by the US Department of Agriculture is considered the most coherent and stringent. It is defined as 'a system that is designed and maintained to produce agricultural products by the use of methods and substances that maintain the integrity of organic agricultural products until they reach the consumer. This is accomplished by using substances, to fulfill any specific fluctuation within the system so as to maintain long term soil biological activity, ensure effective peak management, recycle wastes to return nutrients to the land, provide attentive care for farrn animals and handle the agricultural products without the use of extraneous synthetic additives or processing in accordance with the act and the regulations in this part'. The origin of farming goes back, in its recent history, to 1940s. During this period, the path breaking literature on the subject published by J.I. Rodale in the United States, Lady Balfour in England and Sir Albert Howard in India contributed to the cause of farming. The farming being practiced for the last three decades in India has increasingly been found non-sustainable. The system is oriented towards high production without much concern for ecology and the very existence of man himself.

NEED

There are three categories of opinions about the relevance of farming for India. The first one simply dismisses it as a fad or craze. The second category, which includes many farmers and scientists, opines that there are merits in the farming but we should proceed cautiously considering the national needs and conditions in which Indian agriculture functions. They are fully aware of the environmental problems created by the conventional farming. But many of them believe that yields are lower in organic cultivation during the initial period and also the cost of labour tends to increase therein. The third one is all for farming and advocates its adoption wholeheartedly. They think that tomorrow's ecology is more important than today's conventional farm benefits.

However, among many a major reservation, the profitability of farming vis a vis conventional farming, is the crucial one from the point of view of the Indian farmers, particularly the small and marginal.

Farming involves management of the agro-eco system as autonomous, based on the capacity of the soil in the given local climatic conditions. In spite of the ridicule poured out on farming by many, it has come to stay and is spreading steadily but slowly all over the world. India has been very slow to adopt it but it has made Inroads into our conventional farming system. One advantage we have here is the fact that the farming techniques practiced in this country before the advent of the green revolution were basically eco-lriendly and they have not faded away from the memories of the present elder generation of our farming community. India's options in finding out an alternative method to the conventional farming are limited. The World Trade Organization deadline for banning exports of vegetables and horticulture products based on infarming will expire in 2005. Sensing the importance, the Central and state governments have taken several initiatives to popularise farming in the country.

OBJECTIVES

- To understand the need for farming in India in the light of the experiences of other countries.

- To assess and evaluate the factors which may facilitate the adoption of farming in the country.
- To analyse the constraints, both political and social, and above all economic, in the introduction of farming in India.

RELEVANCE OF FARMING

The relevance and need for an eco-friendly alternative farming system arose from the ill effects of the chemical farming practices adopted w^orldwide during the second half of the last century. The methods of farming evolved and adopted by our forefathers for centuries were less injurious to the environment. People began to think of various alternative farming systems based on the protection of environment which in turn would increase the welfare of the humankind by various ways like clean and healthy foods, an ecology which is condusive to the survival of all the living and non-living things, low use of the non-renewable energy sources, etc. Many systems of farming came out of the efforts of many experts and laymen. However, farming is considered to be the best among all of them because of its scientific approach and wider acceptance all over the world.

THE INTERNATIONAL SCENE

The negative effects of modem chemical based farming system were first experienced by those countries, which introduced it initially. So, naturally, it was in those countries farming was adopted in relatively large scales. There are very large organisations promoting the farming movement in European countries, America and Australia etc. These organisations the International Federation of Organic Agriculture Movements and Greenpeace have studied the problems of the chemical farming methods and compared the benefits accruing to the organic faming with the former. Farming movements have since spread to Asia and Africa too.

IFOAM was founded in France in 1972. It spearheads and coordinates farming efforts the world over by promoting organic agriculture as an environment friendly and sustaining method. It focuses on farming by highlighting the minimum pollution and low use of non-renewable natural resources through this method. It has about 600 organisational members spread over about 120 countries including India.

IFOAM undertakes a wide range of activities related to farming such as exchanging knowledge and thoughts among its members; representation of the movement in governmental, administrative and policy making forums in the national and international arena; updating of production, processing and trading standards; formulation and coordination of research projects; and holding of international conferences and seminars. IFOAM participates in the activities related to farming under the auspices of the United Nations and keeps active contacts with several international NGOs.The Food and Agriculture Organisation of the United Nations provides support to farming in the member countries. It also attempts the harmonization of national organic standards, which is absolutely essential to increase international trade in organic products. The FAO has, in association with the World Health Organisation, evolved the Codex Alimentarius for organic products.

Farming has several advantages over the conventional one apart from the protection of both the environment and human health. Improved soil fertility, better water quality, prevention of soil erosion, generation of rural employment, etc. are some of them.

The concept of quality food has undergone a drastic change over the past few decades. It does give emphasis on the characteristics of the end product, but the process and method of production and transport are now considered equally important. Not only the importers but also the domestic retailers have their own quality specifications or standards, which in many cases are tougher than those of the government regulations. Consumers have become health conscious and are willing to pay for the clean, healthy and natural food. Many developed countries have various support programmes to help farming with financial incentives and technical guidance.

The organic food market in the world has grown rapidly in the past decade. International trade in organic foods showed an annual growth rate of about 20-22 per cent during this period. Many retail chains and supermarkets in advanced countries are accorded with 'green status' to sell organic foods. The organic food processing industry is considered nature friendly and thus encouraged. Table presents the size of the organic food markets in some developed countries in 2000. The US food market tops with a value of $8.0 billion, followed by Germany and Japan with $2.5 billion each while France accounts for $1.25 billion. Moreover, the organic foods command higher prices which often range from 10 to 100 per cent more than the conventional ones.

Table. The World Organic Food Market

{In billion US $) (estimates) Sr. No.	**Country**	**1997**	**2000**
1	USA	4.20	8.00

2	Germany	1.80	2.50
3	Japan	1.20	2.50
4	Italy	0.75	1.10
5	France	0.72	1.25
6	UK	0.45	0.90
7	Australia	NA	0.17
8	China	NA	0.12
9	New Zealand	NA	0.58
10	Taiwan	NA	0.10
11	Philippines	NA	0.06
12	Others	1.33	10.38
Total	10.45	19.73	

The important organic products traded in the international market are dried fruits and nuts, processed fruits and vegetables, cocoa, spices, herbs, oil crops and derived products, sweeteners, dried leguminous products, meat, dairy products, alcoholic beverages, processed foods and fruit preparations. Cotton, cut flowers, animals and pot plants are major non-food products in the world markets.

Farming is bound to grow around the world as many countries are developing their own standards and regulations. The US and the European Union have very comprehensive National Organic Programmes and the early nineties have seen farming regulations in Japan, Canada and Australia. New Zealand, Israel and Brazil have adopted the organic standards equivalent to those of USA and the European Union. China, Thailand, South Korea, Philippines, Turkey and Mexico have established certifying agencies. India too has adopted the National Programme for Organic Production with national standards.

There were about 35,000 certified organic farms in 17 countries during 1998 covering an area of more than a million hectares. A survey held recently revealed that 6-7 per cent farms in USA and Germany are certified organic farms. Many organic inspection and certifying agencies like Crop Improvement Association, IFOAM, Ecocert, BCS, SKAL. KRAV, IMO, FVO, QAI, Oregon, Tilth and NASAA have strong international presence and have even developed the ability to assist marketing of organic products world wide.

Certification of products is the foundation on which organic agriculture rests. Its role in production and marketing of organic foods becomes a significant corner stone in the creation of a confident consumer community.

The need for a trustworthy and healthy organic market is important as the consumers are becoming increasingly separated from the food production process. The void is filled in by a plethora of processing, distribution and marketing intermediaries. The consumers should possess knowledge and confidence that the products they buy are truly organic. Such environment can be created by an organised system of inspection and certification supported by law and regulations of the state. Organic products enjoy trust based on the label or logo approved by the certification process.

The aim of organic certification is, therefore, to:

- Provide an identifiable label (a logo or a symbol); and
- Give an assurance to the consumers that the product is truly organic.

It is widely recognised the world over that the certification of organic products should be based on the following principles:

- Organic production and processing standards should be clearly laid down.
- The conformation of production and processing to these stands must be verified.
- Organic labels should be permitted only to those produces, which are found conforming to the set standards.

Thus, a label on an organic product conveys that the manufacturer has a license for organic production, an independent agency has inspected the production/processing practices followed by the producer, and compliance of the set standards is made. Such an assurance becomes crucial in the generation of consumer confidence in the organic products, particularly as they are costlier than the conventional ones. The marketing agents of organic products also need such an assurance to maintain fair trade practices essential to the growth and sustainability of the demand for such products.

Growth of Farming

Farming has spread to about 100 countries around the world. An estimate in 2004 puts about an area of 24 million hectares under oi'ganic farming worldwide. Australia with its 10.5 million hectares leads the countries.

However, much of this area is pastoral land for grazing. Argentina with 3.19 million hectares, Italy with 1.83 million hectares and USA with 0.95 million hectares follow suit. The importance of farming is growing in many countries. Austria and Switzerland have about 10 per cent of their food system under organic agriculture. The annual growth of farming is estimated to be about 20 per cent in USA, France, Japan and Singapore.

The Developed World

Table shows the top 15 countries in terms of area under farming, percentage of organic area to the cultivated area the number of organic farms :

Table. Area under Farming and Number of Farms Sr.

No.	Country	Area under Organic Farming (ha)	Percentage of area to cultivated area	Number of Organic farms
1	Australia	10500000	2.31	1380
2	Argentina	3192000	1.89	1900
3	Italy	1830000	7.94	56440
4	USA	950000	0.23	6949
5	Uruguay	687481	4.00	334
6	United Kingdom	679631	3.96	3981
7	Germany	632165	3.70	14703
8	Spain	485079	1.88	15607
9	Canada	430600	0.58	3236
10	France	419750	1.40	10364
11	China	301295	0.06	2910
12	Brazil	275576	0.08	14866
13	Chile	273000	1.50	300
14	Austria	258500	11.30	18292
15	Czech Republic	218114	5.09	654

Sr. No.	Country	Area under Organic Farming (ha)	Percentag e of area to cultivated area	Number of Organic farms
1	Liechtenstein	17.00	690	35
2	Austria	11.30	258500	18292
3	Switzerland	9.70	102999	6169
4	Italy	7.94	1830000	56440
5	Finland	6.60	147943	4983
6	Denmark	6.51	174600	3525
7	Sweden	6.30	193611	3589
8	Czech Republic	5.09	218114	654
9	Uruguay	4.00	687481	334
10	United Kingdom	3.96	679631	3981
11	Germany	3.70	632165	14703
12	Norway	2.62	26673	2099
13	Slovakia	2.40	58706	82
14	Australia	2.31	10500000	1380
15	Eastland	2.00	20141	369

Australia had the largest area under farming and Argentina was the next, followed by Italy. However, in terms of percentage of farming area to that of the cultivated, Australia was the fourteenth and Italy fourth. In terms of number of farms under organic agriculture, Italy topped. An examination of all these three characteristics, indicates that Italy, Germany and Austria are the leading countries in organic agriculture and they are among the developed nations of the world. Not even a single developing country except Argentina in terms of area found a position among the top fifteen described in the Table. However, it should be remembered that the US was the fourth in terms of area under farming and had the largest market for organic products in the world.

USA

Farming did spread fast in USA during the 1990s. The growth was in response to the rising demand for organic products both in the national and international markets. The US agriculture farms view farming as a method to reduce input costs and use of non-renewable resources, which increases accessibility to the high value markets and assures premium prices for the produces and above all increased farm income. The US farming system uses eco-friendly agriculture practices in cultural and pest management and avoids the application of chemicals. Antibiotics and hormones are not used in livestock production.

The growth of consumer demand for organic products has also resulted in the specialization in growing, processing and marketing of these products. The implementation of national organic standards in 2002 by the USDA has further boosted the demand for organic products in the country.

Because of increasing prices and ineffectiveness of chemicals, many farms in USA do not favour intensive chemical based agi-iculture. A survey of certified organic farms revealed that more than 60 per cent of them grew vegetables and they had livestock for manure to use in the farm. The farmers had identified three problem areas, which required study.

They were:

1. The need for increasing demand for organic products,
2. The influence of cultural practices on quality and nutritional value of the products; and
3. The relationship between plant nutrition and resistance to pests.

In another study of consumer demand for organic foods, it was found that about 30 per cent of shoppers used some organic food in a year, while 10 per cent used it once a week. The value of sales of organic products in US was $1.5 billion in 1992 and $8 billion in 2000. Many methods to promote sales and consumption of organic food have been introduced. The US Senate approved a bill to make a part of the Federal School Lunch Programme organic. Organic food shows are held annually to give awards for the best organic foods and drinks. The Federal government encourages the planting of native plants for Municipal planting programmes. The government has initiated polices to reduce the application of pesticides and labelling food to indicate the amount of pesticides used to produce the commodity is compulsory. There are about 17 organic certification agencies in US.

The certification and labelling laws are very rigid in United States. Imports of organic products must conform to the standards laid down by the USDA and naturally, the ability of many developing countries to comply with these requirements is limited. A grower or a dealer in need of certification has to lodge a plan to an accredited agency of the USDA. There are both private and state certification programmes. The plan must contain all details in relation to the farming or handling methods and the materials those will have to be used therein. The anticipated improvements and future aims should also be a part of the plan. The farmers or dealers of wild crops should also submit a plan mentioning that the harvesting techniques to be employed will not cause any damage both to the produce and the environment.

All records related to the management practices and the materials utilized in organic production should be maintained for five years. Certification will be approved only if the crops are grown on the land free of prohibited substances for three years prior to harvest. The period of conversion is three years between the conventional and farming and the crops grown during this period cannot be labelled organic or any other label indicating that they are under conversion.

The Organic Foods Production Act contains a wide and detailed coverage of farming methods and materials along with the soil fertility, application of manure, crop rotation and composting. The National Organic Standards Board's recommended ingredients include crop residues, crop wastes from food processing operations, animal manure, yard wastes from private and municipal sources, or other vegetable by-products. Municipal solid waste compost and sewage sludge composts and the use of prohibited material as composts/ ingredients are banned. There is a National List of acceptable and banned materials which includes pest control treatments and other agricultural inputs like fertilizers and seed treatments. An evaluation of all agricultural inputs has been made for their long term effect on the environment irrespective of the fact that they are synthetic or natural.

The exporters of other countries to the US market will have to face these very tough specifications, particularly if their national standards are not equivalent to those of the USDA.

Italy

The European Union of 25 countries had 5.5 million hectares under farming at the end of 2002 managed by 1,60,000 organic farms. This constitutes about 3.4 per cent of the agricultural area and 1.7 per cent of farms in EU. It is estimated that the growth in the area was 9 per cent over the last year. The increases came mainly from France, Spain and the United Kingdom. The European Commission has formulated the 'European Action Plan for Organic Food and Farming' on 10* June 2004. It proposes 21 specific actions to promote farming in the EU countries and the Action Plan is with the European Parliament for consideration. As said earlier, the progress achieved by Italy, Austria and Germany was impressive and as such it is discussed here. The pioneers of farming in Italy had contributed immensely to the spread of organic agriculture, which is acclaimed to be one of the best in the world. Farmers, medical doctors, food processors, traders and consumers were in the forefront working hard from the 1950s for the cause of farming in the country.

Growth of farming in Italy has been spectacular during the period of 1990-2000. The organically cultivated area increased from 13,000 ha to more than a million ha which came to about 7.2 per cent of the total cultivated area. The number of farmers was about 13,00 in 1990, which shot up to 47,357 at the end of 2000. At the end of 2002, the area increased to 1,83,000 ha and the number of farms to 56,440. The reasons for such a substantial increase were the food scandals like the mad cow disease leading to the search for safe foods, the constant endeavor for alternative farming methods which were environmentally, technically and economically attractive to the farmers and the liberal subsidies.

The farmers who opted for organic methods prior to 1990 were people of non-agriculture background, people searching for novel life styles, etc. A significant proportion of them consisted of non-Italians like, Germans, Swiss, Austrians and Americans. They had the unity of purpose in establishing cultural associations in cooperation with like minded people by setting up marketing links with organic farms. However, the decade of 1990-2000 saw that the new entrants into farming were Italians and many of them did not have experience of the new methods. But they possessed a good knowledge of the ecology and understood the need for a marketing system. The period also witnessed a change in the geographical distribution of area under organic cultivation. Sardinia became an important region in farming with the extension of subsidies to pasture lands used for sheep rearing and milk production. Sardina and Sicily account for about 44 per cent organic area in Italy.

At present 41 per cent of the organic area is under pastures. Cereal crops are raised on 18.2 per cent of the area and the olive cultivation alone has the highest single crop share of 10 per cent. The high growth of organic production led to the growth of the processing units to the extent of 142 per cent during 1997-2000 alone. Presently, there are about 42,000 processing units of which 1,300 are established by the farmers themselves. Farming has resulted in the setting up, conversion or expansion of new economic activities in the rural areas of Italy.

The first organic shop had come up in Milan, the industrial and financial capital of Italy, during the 1970s under a consumers' cooperative venture. The most important fact of the market distribution of organic products in Italy is the predominant position of the farmers' or consumers' cooperatives. The distribution was mostly done during the immediate post 1970s by small grocery stores and the major share was that of the direct sales in the open market places and at the farms. After 1990, specialized markets, supermarkets, franchisees, school canteens, etc. became the major channels. Supermarkets are entering the field in large numbers since 1999 with their own labels to convince the consumer of the reliability of the products in the midst of numerous labels. Diary products have a share of 26 per cent of the consumer expenditure on organic products; fruits and vegetables accounts for 13 per cent; breads and biscuits had a share of 12 percent. A growing number of kindergartens, school and university canteens, restaurants and factory Ccinteens are opting for organic foods in the country. A survey of the motives of the consumers of organic foods revealed that every one was going for organic products to protect the health of his/her family.

The support for farming by the European Union by various legislations providing subsidies was instruinental in the growth of farming in Italy. The devolution of power to the Regional Government on agriculture policy in 1972 has resulted in different levels of subsidies in different regions/crops.

There are nine organic certification bodies in Italy. Some of them were earlier cultural associations promoting farming. A single agency named ICEA has a share of 24 per cent of the certified farms coming to about 29 per cent of the area. These nine agencies have about 900 local offices, 1000 staff and technicians for inspecting farms, processing units and storages for giving certifications.

Universities and Public Research Centres are becoming increasingly interested in research and extension activities in relation to organic agriculture. A National School of Farming has been established by the Ministry of Agriculture to impart education.

Austria

Austria is a leading country in Europe in the adoption of farming. At the end of 2003, about 11.6 per cent of the cultivated area which came to 2,96,154 ha was under organic agriculture. The number of organic farms came to 18,576, which was 9.2 per cent of the total number of agriculture farms in Austria.

The conversion of cultivated land to organic method of agriculture was slow till 1990. The federal government introduced subsidies for organic conversion of cultivation in 1991. The government programme of agri-environment of 1995 and the conversion of grass lands into organic farms gave a great fillip to the organic movement. There was a decrease in the conversions in 2000, but it increased substantially from 2001 to 2003. The decline was due to the difficulties in marketing the organic produces in the country. This was an indication that the emphasis on farm production was not enough to promote organic agriculture.

The increasing phase from 2001 to 2003 saw about 700 farms having an area of 30,000 ha being converted. This came from lands cultivating cereals and vegetables. One of the reasons for the change is attributed to the realization of the benefits of farming by the young farmers with a better concern for ecology and understanding of economics than their parents. The organic farmers belonged to a number of associations and the latter had different approaches/ objectives, regional differences and marketing techniques. About 64.5 per cent of the organic farmers were members of these associations, which functioned under two umbrella organizations, named ARGE and OelG. These associations have their own production and quality standards, which in many cases are more stringent, than those of the State. Presently there are 11 major farming associations in Austria and while 6 of them come under ARGE, the remaining five and 4 marketing associations are members of OelG, the latter being the one giving importance of marketing too in the promotion of organic agriculture in Austria.

The concentration of organic farms is in the regions with large areas of grass land. It is easier for the conversion of the grass land to farming as they used to be cultivated extensively. Grass land farms constitute about 66 per cent of the total organic farms. The average size of an organic farm in Austria is 16 ha. Certification of organic farms in Austria is done with the provincial government and a contract between the farm and an inspection agency is essential for registration. An approved organic label is available to guarantee the quality of the organic products. The conditions to be fulfilled by an inspection agency for accreditation are very tough in Austria. 17,500 registered organic farms receive government subsidies and about 50 per cent of it is born by the European Union and balance by the federal and provincial governments.

The market value of Austrian organic products was 2.18 billion Euro in 2000 which was 3 per cent of the total food market. About 70 per cent of these products is distributed by the conventional food chains. Now, the distribution of organic foods through the supermarket chains is gaining momentum,. It is reported that the supermarkets selling organic products enjoy an improved image among their customers. Organic farms' cooperatives also market organic products. Cooperative ventures between the organic farmers and hospitals, nursing homes, etc. have come up to sell organic products. Home delivery services and organic shops are also established, particularly in large cities. Agricultural Schools and Colleges offer training programmes/ courses in farming and promote scientific research. The Institute for Farming at the University of Vienna is a major centre for research in organic agriculture.

Developmient of a strategy for co-operation among the organic farmers, their associations, processors and traders is considered a major challenge in the promotion of farming in Austria.

Germany

Germany, where the IFOAM is located at Tholey-Theley since 1987, is a leading nation in the adoption of farming in Europe. The IFOAM has its Regional Group founded in 1991 with the German speaking nations as members. Germany has about 3 per cent of its total agricultural area under farming against 2 per cent of the European countries as a whole. The country has also a large number of processors and importers of organic produces. So, it would be useful to examine the farming movement in that country in some detail.

The history and development of organic agriculture in Germany begins from the days of bio-dynamic farming initiated in 1924. Bio-dynamic theory of farming treats farm as a living being which responds to non-material influences, called dynamic forces. The proponents of bio-dynamic farming believe that these forces can be enhanced by bio-dynamic preparations. Even certification for bio-dynamic agriculture started in 1954 in the country. Organic-bio-dynamic farming was developed in Switzerland in the early twentieth century. Hans Mueller and his wife Maria Mueller were the spirits behind this movement in Switzerland. Mueller had developed the method during the 1950s and the theoretical basis was provided by Hans-Peter Rusch, the German microbiologist and medical doctor.

The damage inflicted to the environment by the modern farming system was clear by the end of 1960s. The ill effects of pollution were obvious and a producer organisation, named Bioland was established in 1971 in Germany. Another organisation for the organic cultivation of fruits and vegetables had been founded in 1961. A federation for farming associations was established in 1988 consisting of six producer organisations. A common basic standard for organic agriculture was laid down in 1984 as a framework for individual organization.

Farming spread very fast in Germany during the subsequent period. State funding for organic agriculture was available from 1989. The unification of both the Germanys in 1990 saw fast growth of farming in the former East German areas. The trade sector felt the absence of a German logo for organic products and the seal 'Biosiegel' was launched by the federal government.

The German government has put the goal of bringing 20 per cent of the agricultural land under farming by 2010 from the 3 per cent of 2002. A series of measures were introduced to achieve the goal. These steps are under the EU programme of support to organic agriculture and the federal programme for farming.

There are eight organic producer organisations in Germany and they represent about 60 per cent of the farming community. All these associations have the common organic seals for labelling the certified products, both primary and processed. The organic producers, processors and traders of the countiy have formed a common organization aiming at the propagation of achievements of organic production, protection of environment etc. There are several other organisations functioning in relation to farming in Germany. SOL, FiBL,

Schweisl\irh Foundation, Foundation Future Agriculture and Society for Ecological Animal Husbandry are some of the institutions contributing to the spread of farming in Germany.

Farming has gained acceptance in the former East Germany compared to the West. The reasons are the substantial changes in the German agriculture after the reunification, existence of large conservation areas in which farming can well be fitted into. Large areas in the eastern part are designated as disadvantaged and incentives are provided there to go for farming. The arguments for and against farming between the organic and conventional farmers are intense in the western parts of Germany and this has to some extent slowed down the spread of the organic agriculture there. However, government regulations and the support of European Union Regulations have made the organic agriculture accepted in the country.

The standards and certification process for farming are stricter in Germany than in other European Union countries. The German regulations restrict certification to the whole farm and the processing standards are tough. There are private and state standards and the inspections are done according to both of them. The use of genetically engineered organisms and techniques are banned in organic agriculture. There are more than 20 inspection bodies and the 22 states of the country control them. Table gives the estimated organic production in Germany.

Table. Organic Production in Germany

Sr. No.	Description	Organic Production	Total Agricultural Production	Percentage of Organic Production to Total
1	Cereals	350,000	45,271,000	0.77
2	Beans	8,000	95,736	8.34
3	Fodder Peas	20,000	610,039	3.28
4	Potatoes	112,500	11,568,000	0.97
5	Sugar Beet	13,500	3,450,000	0.39
6	Oil seeds	22,000	4,368,000	0.50
7	Vegetables	120,098	2,910,000	4.13
8	Fruits	52,338	1,331,000	3.93
9	Beef	36,000	1,438,600	2.50
10	Pork	10,000	397,800	0.25
11	Sheep & Goat Meat	5,500	44,300	12.42
12	Poultry	2,600	825,800	0.31
13	Eggs	150 million	14,342 million	1.05
14	Milk	335	28,334,000	1.18

The German organic food market was worth about 3 billion Euro in 2000. Both the federal and state governments extend financial support to farming. Farmers are encouraged to adopt organic methods from 1989 and those who adopt production methods compatible with the protection of environment are given assistance. In addition to the direct subsidies, marketing assistance is also provided to the organic farms. Subsidies for producers' marketing organisations, processing and market promotion are available.Advisory services are provided to the farmers for farming. Advisory services were organised by the farmers themselves during the initial period of the introduction of farming in Germany. Now producer associations with governmental financial support offer such services. Farmers themselves hire advisors with state support and official advisors are also available.

Research and training facilities are existing in almost all the German agricultural universities and technical colleges in farming. Degree courses are offered in organic agriculture and farming research is also funded by the state. A state research institute for farming was established in 2000.

The Developing World

Large sections of farmers in the developing countries are poor and their land holdings in many countries are small. Their access to external inputs is limited and their ability to improve production is low given the economic and social disabilities to which they are exposed.

Globalization and the opening up of the trade barriers among the nations have resulted in the decline of agricultural prices in the local markets to the detriment of the interests of these small farmers.

These conditions can be made a bit favourable for the introduction of farming in which the traditional system is found conducive for such adaptations with the state support for conversion to organic agriculture. Table shows the area under farming, its percentage to the total and the number of organic farms in the 15 countries which lag behind others in the adoption of organic methods of cultivation.

Table. Countries Lagging Behind in Organic Agriculture

Sr. of farms	Country	Area under farming (ha)	Percentage area to	NumberNoorganicof organic cultivated area
1	Vietnam	2	-	38
2	Zimbabwe	40	-	10
3	Nepal	45	-	26
4	Cyprus	52	0.04	15
5	Syria	74	-	1
6	Benin	81	-	119
7	Croatia	120	-	18
8	Malaysia	131	-	27
9	Laos	150	0.01	-
10	Mauritius	175	0.15	3
11	Fiji	200	0.04	10
12	Jamaica	205	0.04	7

13	Lebanon	250	0.07	17
14	Suriname	250	0.28	-
15	Malawi	298	0.01	6

Sr. organic	**Country organic cultivated**	**Percentage of farms (ha)**	**Area under area to**	**Number farming area**	**No. of**
1	Laos	0.01	150	-	
2	Malawi	0.01	298	6	
3	Tanzania	0.01	5155	991	
4	Guyana	0.02	425	26	
5	Philippines	0.02	2000	500	
6	Thailand	0.02	3429	940	
7	Zambia	0.02	5688	72	
8	Cameroon	0.03	2500	-	
9	India	0.03	41000	5661	
10	Cyprus	0.04	52	15	
11	Fiji	0.04	200	10	
12	Jamaica	0.04	205	7	
13	Ghana	0.04	5453	-	
14	Rep of Korea	0.05	902	1237	
15	South Africa	0.05	45000	250	

Sr. farms	**Country of organic**	**Number organic**	**Percentage**	**Area under area to cultivated area**	**No. of farming (ha)**
1	Syria	1	-	74	
2	Mauritius	3	0.15	175	
3	Malawi	6	0.01	298	
4	Jamaica	7	0.04	205	
5	Fiji	10	0.04	200	
6	Zimbabwe	10	-	40	
7	Cyprus	15	0.04	52	
8	Lebanon	17	0.07	250	
9	Croatia	18	-	120	
10	Guyana	26	0.02	425	
11	Nepal	26	-	45	
12	Island	27	0.24	5466	
13	Malaysia	27	-	131	
14	Ukraine	31	0.40	164499	
15	Liechtenstein	35	17.00	690	

Most of the countries lagging behind in the adoption of organic agriculture were from Asia and Africa. In terms of both in absolute and percentage area under farming, Tanzania, Zambia, Ghana, India and South Africa are far behind many nations in organic agriculture.

NEED FOR FARMING IN INDIA

The need for farming in India arises from the unsustaina-bility of agriculture production and the damage caused to ecology through the conventional farming practices.

The present system of agriculture which we call 'conventional' and practiced the world over evolved in the western nations as a product of their socio-economic environment which promoted an over riding quest for accumulation of wealth. This method of farming adopted by other countries is inherently self destructive and unsustainable.

The modern farming is highly perfected by the Americans who dispossessed the natives of their farms right from the early period of the new settlers in US. The large farms appropriated by the immigrants required machines to do the large scale cultural operations. These machines needed large amount of fossil fuels besides forcing the farmers to raise the same crops again and again, in order to utilize these machines to their optimum capacities. The result was the reduction of bio-diversity and labour. The high cost of the machines necessitated high profits, which in turn put pressure to raise productivity. Then, only those crops with high productivity were cultivated which needed increased quantities of fertilizers and pesticides. Increasing use of pesticides resulted in the damage to environment and increased resistance of insects to them. Pesticides harmed useful organisms in the soil. The monoculture of high yielding seeds required external inputs of chemical fertilizers. The fertilizers also destroy soil organisms. They damage the rhizobia that fix nitrogen and other micro organisms that make phosphates available to plants. The long term effect was reduction of crop yields. The damaged soil was easily eroded by wind and water. The eroding soil needed use of continuously increasing quantities of fertilizers, much of which was washed/ leached into surface and underground water sources.

The theme of consumer welfare has become central in the economic activities in the developed countries in the world. Sustainable agriculture based on technologies that combine increased production with improved environmental protection has been accepted as absolutely essential for the maximization of the consumer welfare. The consumers are increasingly concerned about the quality of the products they consume and food safety has become a crucial requirement. Safety, quality and hygienic standards are increasingly being made strict. The mad cow disease and the question of genetically modified food production are the recent instances, which made the countries to tighten the laws. Mycotoxln contamination, unacceptable levels of pesticide residues and environment degradation are the problems on which the attention is centred. Keeping the interests of the consumers, the European Union has taken tough measures including criminal prosecution to ensure food safety. Another area to increase the consumer welfare is promotion of the eco-friendly methods in agriculture. No-till, or conservation agriculture, lower input approaches of integrated pest or nutrient management and farming are some of them.

The Indian agriculture switched over to the conventional system of production on the advent of the green revolution in the 1970s. The change was in the national interest which suffered set backs because of the country's over dependence on the foreign food sources. The national determination was so intense that all the attention was focused on the increase in agriculture production.

The agriculture and allied sectors in India provide employment to 65 per cent of the workers and accounts for 30 per cent of the national income. The growth of population and the increase in income will lead to a rise in demand for foodgrains as also for the agricultural raw materials for industry in the future. The area under cultivation, obviously, cannot be increased and the present 140 million hectares will have to meet the future increases in such demands. There is a strong reason for even a decline in the cultivated area because of the urbanization and industrialization, which in turn will exert much pressure on the existing, cropped area.

Science and technology have helped man to increase agricultural production from the natural resources like land. But the realization that this has been achieved at the cost of the nature and environment, which support the human life itself, is becoming clear. It has been fully evident that the present pattern of economic development, which ignores the ecology and environment, cannot sustain the achievement of man without substantial erosion of the factors that support the life system of all living things on the Earth. The evidence of the ill effects of development is well documented. As said earlier, we in India have to be concerned much more than any other nation of the world as agriculture is the source of livelihood of more than 6-7 million of our people and it is the foundation of the economic development of the country.

There were times when people lived close to nature with access to flora and fauna in healthier and cleaner surroundings. One has to look back at our present metropolitan cities or other large towns before the past fifty years as recorded in history/memmories of the present elder generation to see the striking differences in the surroundings in which the people lived there. Land, water and air, the most fundamental resources supporting the human life, have degraded into such an extent that they now constitute a threat to the livelihood of millions of people in the countiy.

Ecological and environmental effects have been highly publicised all over the world. Many times, these analysis have taken the shape of doomsday forecasts. Powerful interests in the developed western countries have also politicised these issues to take advantage of the poor nations of the world. Efforts to impose trade restrictions on the plea of environment protection are a direct result of these campaigns. But we have to recognize that the abysmal level to which we have degraded our resources ,requires immediate remedial measures without terming the demand for them as the ploys of the rich nations to exploit the poor.

Another turn of the events has been the blame game for ecological problems stated at the Earth Summit and other international conferences. The developed countries, it is true, are to a great extent instrumental to degrade the environment. However, the poorer countries of the world including India cannot delay or ignore the need for remedial measures, which are to be effectively implemented. We cannot gloss over the fact that we have also contributed to the degradation of ecology; look at the droughts and floods, disappearance of forests, high noise level and air pollution in the cities which are our own creations.

Organically cultivated soils are relatively better attuned to withstand water stress and nutrient loss. Their potential to counter soil degradation is high and several experiments in arid areas reveal that farming may help to combat desertification. It is reported that about 70 hectares of desert in Egypt could be converted into fertile soil supporting livestock through organic and biodynamic practices. India, which has some areas of semi-arid and arid nature, can benefit from the experiment. The organic agriculture movement in India received inspiration and assistance from IFOAM which has about 600 organizational members from 120 countries. All India Federation of Farming is a member of IFOAM and consists of a number of NGOs, farmers' organisations, promotional bodies and institutions.

The national productivity of many of the cereal crops, millets, oilseeds, pulses and horticultural crops continues to be one of the lowest in the world in spite of the green revolution. The fertilizer and pesticide consumption has increased manifold; but this trend has not been reflected in the crop productivity to that extent. The country's farming sector has started showing indications of reversing the rising productivity as against the increasing trend of input use.

The unsustainability of Indian agriculture is caused by the modern farming methods which have badly affected/damaged production resources and the environment.

Affects of Modern Farming Technology

Affects of Modern Farming Technology

The role of agriculture in economic development in an agrarian country like India is a pre-dominant one. Agriculture provides food for more than 1 billion people and yields raw materials for agro-based industries. Agricultural exports earn foreign exchange. Modernization of Indian agriculture began during the mid-sixties which resulted in the green revolution making the country a foodgrain surplus nation from a deficit one depending on food imports. Modern agriculture is based on the use of high yielding varieties of seeds, chemical fertilizers, irrigation water, pesticides, etc., and also on the adoption of multiple cropping systems with the extension of area under cultivation. But it also put severe pressure on natural resources like, land and water. However, given the continuous growth of modern technology along with the intensive use of natural resources, many of them of Non- renewable, it is felt that agriculture cannot be sustainable in future because of the adverse changes being caused to the environment and the ecosystem. The environmental non-degradable nature of the agricultural development and its ecological balance have been studied in relation to the modem Indian farming system by experts which shows exploitation of land and water for agriculture, and the excessive use of chemicals.

Chemical Contamination

Fertilizers

Consumption of chemical fertilizers {N,P,K) has been increasing in India during the past thirty years at a rate of almost half a million tonnes on an average, a year. It was only 13.13 kg/ha in 1970-71, 31.83 kg/ha in 1980-81 and 74.81 kg/ha in 1995-96. It shot up to about 96 kg/ha during 1999-2000. Table shows the consumption of fertilizers in India from 1970-71 to 2001-02.

Table. Consumption of chamical fertilizers in India.

Sr.	Year	Consumption	Consumption (Kg/ha)	No.
(M. Tonnes)				
1	1970-71	2.18	13.13	
2	1980-81	5.52	31.83	
3	1990-91	12.54	67.49	
4	1991-92	12.73	69.84	
5	1992-93	12.15	65.53	
6	1993-94	12.24	66.69	
7	1994-95	13.56	73.12	
8	1995-96	13.58	74.81	
9	1996-97	14.31	76.70	
10	1997-98	16.19	86.80	
11	1998-99	16.80	89.80	
12	1999-00	18.07	95.60	
13	2000-01	16.71	NA	
14	2001-02	17.54	NA	

The present use of about 96 kg of fertilizers per ha in India appears to be modest compared to the advanced countries. Currently about 80 per cent of the fertilizer is consumed in only about 120 districts constituting less than 33 per cent of the gross cultivating area. Experts point out that the efficiency of fertilizer use in India is only 30-35 per cent as the balance 65-70 per cent reaches the under ground water. The intensity of their use in a few regions and a few crops are causes of serious concern to human health, soil, water, environment and thus to the sustainability of agriculture production in the country.

It is true that the increasing use of fertilizer at high rates has boosted agricultural production in the country. But it has also caused adverse impact on soil and water as well as environment. Several studies on the effects of high level of fertilizer application on soil health have confirmed the adverse impacts.

Both drinking and irrigation water wells in large numbers have been found contaminated with nitrates, some of them are having even 45 mg per litre, well above the safe level.

Long term continuous use of high doses of chemical fertilizers badly affects the physical, chemical and biological properties of the soil. A study at the University of Agricultural Sciences, Bangalore confirmed the deterioration of soil health because of the reduction in water holding capacity, soil pH, organic carbon content and the availability of trace elements such as zinc in case of ragi crop even with the application of normal doses of fertilizer in the long run. In the long run, increasing nitrogenous fertilizer use leads to the accumulation of nitrates in the soil. The application of sulphatic fertilizers leaves sulphates in the soil. Rainfall and excessive use of irrigation water cause these chemicals to change the alkaline or acidic nature of the soil. The nitrates go to the rivers, wells, lakes etc. And also leak into the drainage system which goes into the drinking water contaminating the environment. It also causes depletion of the ozone layer adding to the global warming. Use of nitrogen in the form of ammonium sulphate in the rice crop emanates ammonia polluting the atmosphere. The heavy metals present in the fertilizers and sewage sludge leach into ground water. Table shows the content of some heavy metals in fertilizers and sludges.

Table. Content of heavy metals in fertilizers and sludges

(Matal: mg/kg)

Source	**Cd**	**Cr**	**Cu**	**Pb**	**Zn**
Ammonium nitrate	1.1	2.5	3.6	5.4	11.7
Super phosphate	16.6	157.0	22.6	20.6	244.0
NPK fertilizers	4.9	54.3	8.3	3.2	97.5
(8-10-18)					Sewage sludge 20.0
500.0	250.0	700.0	3000.0		

Pesticides

The use of chemical pesticides began with the discovery of toxicological properties of DDT and HCH during the Second World War. Many chlorinated hydrocarbon insecticides like aldrin, dieldrin, toxaphane, chlordane, endosulfan, etc. came into the market during the second half of the last century. Simultaneously, organophosphate and carbonate compounds were employed in agriculture. A new group of insecticides, such as premethrin, cypermethrin, fenalerate, etc. which were effective at low doses came into being in the 1970s. The use of pesticides has helped in increasing agriculture production and also led to the development of resistance in pests, contamination of the environment and resurgence of many pests.

There are about 1000 agrochemicals in use in the world over. India accounts for about 3.7 per cent of the total world consumption. At present, our consumption is about 90,000 tonnes of plant protection chemicals. It comes to about 500 grams per ha compared to 10-12 kg/ha in Japan and 5 kg/ha in Europe. However, the use of pesticides in India is uneven like the fertilizers. While in cotton it is about 3 to 4 kg/ha, in pulses it comes to below 500 grams/ha.

Agricultural chemicals have become a major input in Indian agriculture with the increasing demand for food, feed and fibre. The pesticide consumption was about 2000 tonnes annually during the 1950s. India happens to be the second largest manufacturer of pesticides in Asia after Japan. It is also of interest to know that in spite of increased consumption of plant protection chemicals, the produce loss due to insects and pests increased by 5 times during the period from 1988 to 1995.

Increasing application of fertilizer also leads to increasing use of pesticides to control pests and diseases. The trend of increasing fertilizer use also compels the farmers to enhance the use of pesticides as well. the use of fertilizers in increasing amounts leads to growth of weeds and in the process of weedicide use many plants growing nearby also get killed, which reduces the biodiversity. Meanwhile, the weeds also develop resistance to herbicides and the quest to formulate even powerful herbicides begins.

Pesticide consumption in India from 1970-71 to 2001-02 is shown in Table.

Table. Pesticide consumption in India.

Sr. No.	Year	Consumption (.000 tonnes)
1	1970-71	24.32
2	1980-81	45.00
3	1990-91	75.00
4	1991-92	72.13
5	1992-93	70.79
6	1993-94	63.65
7	1994-95	61.36
8	1995-96	61.26
9	1196-97	61.26
10	1997-98	56.11
11	1998-99	52.44
12	1999-00	49.16
13	2000-01	46.20
14	2001-02	44.58

Consumption of pesticides increased from 24.32 thousand tonnes in 1970-71 to 75 thousand tonnes in 1990-91 and it slowed down during the subsequent period. Insects, pests and diseases like viral, bacterial and fungal affect the high yielding varieties of crops. Almost all pesticides are toxic in nature and pollute the environment leading to grave damage to ecology and human life itself. This indiscriminate use leaves toxic residues in foodgrans, fodder, vegetables, meat, milk, milk products, etc. besides in soil and water.

High doses of pesticides severely affect the aquatic animals, fish and the wild life. Insects develop resistance to insecticides in crops like cotton and in turn force the farmers to the excessive use of them. Cases of pesticide poisoning and human and animal deaths are also reported. Pesticides irritate the skin and the respiratory system in the humans gets damaged.

It was found that all water bodies like, rivers, canals, lakes, tanks and ponds and also the costal water were contaminated with high amounts of DDT, HCH and other organochlorine pesticides. River water is seen as more contaminated than other water sources. Contamination of drinking water with DDT and HCH is reported from different states. Since the concentrations of contaminants are higher than MRL values fixed by the Environmental Protection Agency, the seriousness of the problem can be gauged.

Pesticides also contaminate animal feeds and fodder. Green fodder, paddy and wheat straw contain residues of DDT and HCH. Several studies have confirmed this trend.

Milk and milk products are also affected by the pesticide use. Both bovine and human milk showed high levels of pesticide contamination. The sources of contamination of bovine milk are traced to the fodder and feed concentrates and in case of the human milk, the consumption of contaminated food by the lactating mothers is reported to be the reason.

Infant formula/baby milk powders also showed DDT and HCH contamination level ranging from 94 to 100 per cent. Butter and ghee, the other animal products revealed high contamination levels in many parts of the country.

Cereals like wheat and rice were seen contaminated highly by pesticides like, DDT, HCH and malathion. The case of vegetable, vegetable oil, honey, fish etc. is also not different as they too have unacceptable high pesticide residue content levels.

The adverse effect of pesticide contamination on humans in India is understood from the study of dietary intake. Such studies, although a few in number, have confirmed high levels of pesticides contamination which come to more than 3 to 5 times than the agriculturally developed countries.

The daily intake of pesticide per individual is estimated to be about 0.51 milli-grams which is above the accepted level. The Indian Institute of Horticulture Research has reported contamination of 50 per cent of the fruits and vegetables sold in the Bangalore market with the residues of DDT and HCH.

Use of herbicides over a period results in the shift of the weed flora. The weeds of minor importance, often, become major weeds. Repeated application of weedicides helps the development of resistance in weed at alarming proportions. The remedy recommended is rotation of herbicides or the use of other herbicides. Any way, the end result is contamination of ground water and soils inflicting damage on environment.

The number of herbicides registered in India comes to about 28 in 1997-98 which was only 10 before 10 years. This is often compared to about 300 herbicides available in the North America. There are only 10 herbicides manufactured in the country and the herbicides consumption was about 6000 Tonnes during 1994-95. It is reported that the level of herbicide use in rice, wheat, and tea in India is almost the same that of the world at large. Sugarcane, soyabean, groundnut, coffee, cotton, onion and potato are the other crops, which find widespread application of herbicides in India.

The contamination of water, air and soil with toxic synthetic fertilizers, pesticides and herbicides leads to increasing deaths of many creatures, and to human illness and mortality.

The end result is loss of biodiversity and natural harmony, increased expenditure to purify water, air, etc. The toxins in the food crops cannot be removed and the threat to human existence itself seems to be real.

The firms engaged in the manufacture and supply of agricultural inputs have a vested interest in keeping the input use increasing. Besides, they influence the government policy towards agriculture. '

Salinity and Water Logging

Water is one of the important inputs for the vigorous growth and high yields of crops. The modernisation of Indian agriculture has resulted in the increased use of irrigation water. The area under irrigation has grown substantially during the past three decades. Table shows the gross irrigated area in the country.

Table. Gross area under irrigation in India

Sr. No.	Year	Gross irrigated area (Million ha)
1	1970-71	38.18
2	1980-81	49.73
3	1990-91	62.47
4	1991-92	65.68
5	1992-93	66.76
6	1993-94	68.37
7	1994-95	70.65
8	1995-96	71.35
9	1996-97	73.25
10	1997-98	72.78

The gross irrigated area of 38.18 million ha in 1970-71 increased to 49.73 million ha in 1980-81 and the next decade ending 1990-91 saw this further rising substantially to about 62.47 million ha. It increased to 72.78 million ha during 1997-98.

Heavy irrigation is necessary to get high production, as the new varieties cannot withstand water scarcity. This leads to salinity and water logging leaving the land uncultivable. Over exploitation of underground water is another effect. When water table falls, increasing energy will be required to lift water for irrigation.

Irrigation is necessary for the vigorous growth and high yields of crops in the modern method of cultivation. Many of the crops, particularly the rice and wheat high yielding varieties need more irrigation water than the traditional varieties. The area under irrigation in the country is only about 35 per cent and the remaining is still dependent on rains. So, there is a necessity to use irrigation water judiciously. Its excessive use results in severe ecological dangers like water logging of vast cultivated areas by seepages from canals. The loss of water through seepages and evaporation is estimated to be about 38 per cent. Flooding also results in run off and leaching losses of fertilizer nutrients, pesticides and soil particles. Excessive use of canal water makes the field vulnerable to soil erosion. The excessive irrigation in certain areas results in wastage as evident from the water logging of vast cultivated areas caused by the seepage from many virater sources.

Water logging is harmful to the soil. Seepage of canal water leads to salts present in the lowest layer of soil come up to the surface and the soil may turn alkaline or saline. Dams and multipurpose projects degrade the soil in the command area due to soil salinity and water logging. The chambel region in Rajasthan and Madhya Pradesh, the command area of the Bhakra Nangal, etc are the examples of water logging created by huge water irrigation projects.

Crops irrigated by sewage water have adverse effects on the health of the human population consuming the produce. The workers work on these farms also face health hazards.

Depletion of Energy Resources

Chemical fertilizers, pesticides, herbicides, etc are manufactured using the non-renewable materials like the fossil fuels. The global demand for oil and natural gas is increasing and thus the price of the inputs to agriculture is bound to rise. India's petroleum resources, which presently meet only about 30-35 per cent of the consumption demand, are under pressure. Increasing demand for chemicals and energy in agriculture sector will have affects on our energy sources.

The investments in agriculture have to be increased to meet the rising input costs and larger areas are brought under farming to earn profits. Large farms have to transport the produce to distant areas. Again, energy will be required for transportation, processing and packaging.

The rice-wheat cropping pattern and the cultivation of crops like sugarcane require high irrigation, which results in the depletion of water level. Singh and

Singh found that the water level in the states of Punjab and Haryana had gone down by 0.3 to 1 m per annum during a period of 10 years due to the excessive use of water for paddy crop.

Input-Output Lmbalance

A crop, in its growth process, incorporates a part of the soil fertility into the parts of the plant. The roots remain in the soil. The leaves and stems are fed to the cattle/burnt as fuel/directly returned to the soil. The consumed part by cattle and human also go back to the soil. The practice of commercial farming leads to continuous export of the soil fertility to outside the farming areas as the organic matter leaves the locality. The soil nutrients in the form of farm produces continue to be exported. The import of chemical fertilizers cannot compensate the loss of soil nutrients through exports. The soil becomes powdery and gets eroded by wind or rain. If the harvests are exported from the country, the loss is higher.

The Earth can produce only a limited amount of biomass from a given area. If man tries to extracj: more, the system degenerates. If we realize that we live in a closed eco-system and act in tune, the resources can be used iri a sustainable way. The maintenance of a dynamic equilibrium balances the inputs and outputs in a region. If agriculture becomes unsustainable, no society can survive for a long time.

Expansion of Cultivated Area

Not only the intensive cultivation through the use of technological inputs, but also the extensive crop production through Increase in the area under cultivation has been an important aspect of modern agriculture seen in India. Increasingly areas under forests are brought under plough along with the marginal, sub-marginal and undulating land. The net sown area was 140 million ha in 1970-71 and stood at 142 million ha at the end of 1997-98.

Reduction in Genetic Diversity

The genetic base of crops is very important and a reduction of genetic diversity leads to the emergence of pests on a large scale. Farmers, in olden times, apart from' using the crop rotation methods to maintain the soil fertility also relied on the genetic means to increase crop production. Relying exclusively on nation's own reserves of fertility and immunology, the farming community by evolving trial and error methods discovered. hybrid varieties of crops by crossing the related strains. These crosses were from the same environment and no violence was used to separate them from nature by maintaining the ecological balance. The high yielding varieties of crops are the crosses from different environments and distantly related strains. The high yielding rice variety got by the crossing between the dwarf and non-dwarf varieties has major genetic weaknesses. The dwarf gene is susceptible to pest and viral attacks and the seed cannot manifest its potential without chemical fertilizer. Thus, an artificial environment has to be created for the growth of the crop.

Thus synthetic fertilizers supplant natural fertility, which results in larger population of pests. The new technology adopted then depends upon the replacement of the local/traditional varieties of seeds. But this results in the reduction of genetic diversity and increase in genetic erosion. These modern technologies are but the result of clever manipulations of nature's genes.

Low Productivity

The productivity of cereals, millets, oilseeds, pulses and plantation crops is very low in comparison with those in other countries in the world. This is in spite of our success in improving the quality of seeds and adoption of efficient technology. The impact of green revolution is showing signs of weakness and production appears to have decreased even after an increase in the inputs used. The production of foodgrains in the country increased very substantially during 1960 to 1980 to reach 160 million Tonnes from 60 million tonnes. But the decade ending 1990 and 2000 did not witness such increases and the attainment of the targeted production of 240 million tonnes to meet the demand of the population by 2010 seems to be difficult.

The reasons attributed to the low productivity are the drastic reduction in soil nutrients in the areas where fertilizer is used intensively in which the organic matter is not supplemented.

BENEFITS OF FARMING

Organic agricultural practices are based on a maximum harmonious relationship with nature aiming at the non-destruction of the environment. The developed nations of the world are concerned about the spreading contamination of poisonous chemicals in food, feed, fodder and fibre. Naturally, farming system is looked upon as one of the means to remedy these maladies there. However, the major problem in India is the poor productivity of our soils because of the low level content of the organic matter.The efficiency of the organic inputs in the promotion of productivity depends on the organic contents of the soil. There were many resemblances of farming principles in the traditional agriculture of India. But the former gives a more open and verifiable scientific foundation than the latter.

Healthy Foods

A study conducted in USA on the nutritional values of both organic and conventional foods found that consumption of the former is healthier. Apples, pears, potatoes, corn, wheat and baby foods were analyzed to find out 'bad' elements such as aluminum, cadimum, lead and mercury and also 'good' elements like boron, calcium, iron, magnesium sellenium and zinc. The organic food, in general, had more than 20 per cent less of the bad elements and about 100 per cent more of the good elements.

Improvement in Soil Quality

Soil quality is the foundation on which farming is based. Efforts are directed to build and maintain the soil fertility through the farming practices. Multicropping, crop rotations, organic manures and pesticides, and minimum tillage are the methods employed for the purpose. Natural plant nutrients from green manures, farmyard manures, composts and plant residues build organic content in the soil. It is reported that soil under farming conditions had lower bulk density, higher water holding capacity, higher microbial biomass carbon and nitrogen and higher soil respiration activities compared to the conventional farms. This indicates that sufficiently higher amounts of nutrients are made available to the crops due to enhanced microbial activity under farming. The effect of organic cultivation on soil fertility as reported at the farm of Central Institute for Cotton Research, Nagpur is given in Table.

Table. Organic carbon and available P contents

Year		Organic		Non-organic
Organic	Available	Organic	Available	carbon (per cent)P (kg/ha) carbon (per cent)
P (kg/ha)				
June 1993	0.38	12.1	0.38	12.1
Feb 1994	0.40	12.6	0.36	12.0
Feb 1995	0.46	14.5	0.35	12.9
Feb 1996	0.52	15.0	0.38	12.0

Increased Crop Productivity and Income

Field trials of organic cotton at Nagpur revealed that during the conversion period, cotton yield was low compared to the conventional and integrated crop management. However, the yields of organic cotton started rising from third year. Cotton yields under organic, conventional and the mixed systems were 898, 623 and 710 kg/ha respectively at the end of the fourth year of the cultivation. The yield of soyabean under farming was also the highest compared to the other two systems.

The Central Institute for Cotton Research, Nagpur conducted a study of economics of cotton cultivation in Yavatmal district of Maharashta. The cost of cultivation of cotton was lower in the farming than in the modern system. The low costs were due to the non-use of fertilizers and chemical insecticides. As a result of the low yields during the conversion period, the net income from the organic farm was lesser than the conventional farm. But the yield under organic method increased progressively equalling it to that of the conventional system by the sixth year. The input costs were low under farming and with a 20 per cent of premium prices of output, the net income increased progressively from fourth year under organic fanning. The appreciation of net income from organic cotton cultivation by the sixth year was 80 per cent over the conventional crop. Results reported from 1050 field demonstration cum trials under the National Project on Development and Use of Biofertilizers in different parts of the country show* an increase of 4 per cent in yield in plantation crops, 7 per cent in fruit crops, 9 per cent in wheat and sugarcane, 10 per cent in millet and vegetable, 11 per cent in fibre, condiments and spice crops, 14 per cent in oilseeds and flowers and 15 per cent in tobacco.

A study of 100 farmers in Himachal Pradesh during a period of 3 years found that the total cost of production of maize and wheat was lower under farming and the net income was 2 to 3 times higher. Both productivity and premium prices contributed to the increased profitability. Another study of 100 farmers of organic and conventional methods in five districts of Karnataka indicated that the cost of farming was lower by 80 per cent than that of the conventional one.

Table. The cost benefit ratios described for various crops were

Crops	Organic	Inorganic
Groundnut	1:1.26	1:1.31
Jowar	1:1.36	1:1.28
Cotton	1:1.34	1:1.24
Coconut	1:1.70	1:1.31
Banana	1:3.66	1:2.82

Low Incidence of Pests

The study of the effectiveness of organic cotton cultivation on pests at the farm of Central Institute for Cotton Research, Nagpur revealed that the mean monthly counts of eggs, larva and adults of American BoUworm were far lesser under farming than under the conventional method.

Bio-control methods like the neem based pesticides to Ti-ichoderma are available in the country. Indigenous technological products such as Panchagavya which was experimented at the University of Agricultural Sciences, Bangalore found to control effectively wilt disease in tomato.

Employment Opportunities

Farming requires more labour input than the conventional farming system. Thus, India which has a very large amount of labour unemployment and under employment will find farming an attraction. Moreover, the problem of periodical unemployment will also get mitigated because of the diversification of the crops with their different planting and harvesting schedules resulting in the requirement of a relatively high labour input.

Indirect Benefits

Several indirect benefits from farming are available to both the farmers and consumers. While the consumers get healthy foods with better palatability and taste and nutritive values, the farmers are indirectly benefited from healthy soils and farm production environment. Eco-tourism is increasingly becoming popular and organic farms have turned into such favourite spots in countries like Italy. Protection of the ecosystem, flora, fauna and increased biodiversity and the resulting benefits to all human and living things are great advantages of farming which are yet to be properly accounted for.

PROPOSED OBJECTIVES

The broad objectives of farming in India can be the following in the light of the discussions on the adverse effects of the conventional farming system the country practiced for about 30-40 years and the potential benefits of the organic methods:

- Sustainable agriculture
- Increasing agriculture production
- Food self-sufficiency
- Environmental protection
- Conservation of natural resources
- Rural development

PROGRESS OF FARMING IN INDIA

The first conference of NGOs on farming in India was organized by the Association for Propagation of Indigenous Genetic Resources in October 1984 at Wardha. Several other meetings on farming were held at different places in the country towards the end of 1980s. Here, mention must be made of the Bordi Conference in Maharashtra, the state which was the focal point for the farming movement in India. The Rajasthan College of Agriculture with the support of the state government organized a meeting on organic agriculture in 1992. The United Planters' Association of South India organised two national level conferences on farming in 1993 and 1995. ARISE is a major organization in the country engaged in the promotion of farming. ARISE was founded in 1995 at a national conference of farming held at Auroville. ARISE comprises of a supporting network of regional groups aiming at sustainable environment by protecting bio-diversity and promoting organic agricultural practices. The selection of Auroville for the conference was apt as it housed the Arabindo Ashram and the pioneering work under its auspices on building technology, alternative energy research, wasteland development, afforestation and organic agriculture.

By 1980, three groups of Indians had taken to farming. The first one consisted of urban educated technocrats for peripheral interest, which did not last long. Educated farmers consisted of the second group whose farming practices were based on scientific knowledge. The third group practiced farming through trial and error. The successful organic farmers in India are those who have access to sufficient natural resources like, water and other organic inputs mostly on their own farms. These farms produce crops like sugarcane, areca, cocoa, coconut, pepper and spices. Many of them have shown that switch over to farming do not affect yields and income and more importantly, knowledge/ expertise is available for successful adoption of farming in the country.

The International Federation of Organic Agriculture Movements estimates that an area of about 41,000 hectares in India is under farming representing about 0.17 per cent of the world organic acreage. It also reveals that the percentage of organic area to the total cultivated area comes to only about 0.03 per cent and the total number of farms comes to about 5,661. But, a comparison of our 41,000 ha to Australia, Argentina, Italy and USA clearly indicates that farming in India has to go very far even to catch up with that of the leading nations of the world.

Non Governmental Organizations are spearheading farming in India. A report in 2002 indicates that about 14,000 tonnes of organic products have been raised in India. They include tea, coffee, rice, wheat, pulses, fruits, spices and vegetables. India exports organic agricultural produces to European Union, USA, Canada, Saudi Arabia, UAE, Japan, Singapore and Australia, among others.

The International Conference on "Indian Organic Products-Global Markets" at the end of 2002 was the first to be held in India. IFOAM predicts that India and China have great potential to be organic farm produce exporters in the future. An important event in the history of the modern nascent farming in India was the unveiling of the National Programme for Organic Production on 8"^ May, 2000 and the subsequent Accreditation and Certification Programme on P' October, 2001. The logo "India Organic" was released on 26'*' July 2002 to support the NPOP.

PROGRESS

An important progress towards organic agriculture made by India is the increasing awareness of the ill effects of the modern farming system, which the country adopted about 35 years ago. The threat poised by the conventional food products to the human health and the damage done to the ecology are being viewed seriously. Efforts are made to produce healthy foods and the demand. for them is increasing. The importance of the marketing of the organic products is highlighted for the promotion of organic agriculture. Several individuals and associations have taken to farming and organic products are available in the large cities to a very limited extent.

Production and Exports

The aggregate production of organic agriculture came to about 14,000 tonnes during 2002 and the exports amounted to 11,925 tonnes. Details are given in Table:

Table. Exports of organics products from India-2002

Products	Tonnes
Tea	3000
Rice	2500
Pulses and Vegetables	1800
Cotton	1200
Wheat	1150
Spices	700
Coffee	550
cashew nut	375
Pulses	300
Herbal products	250
oil seeds	100
total	11925

Indian organic products are mainly exported to Europe, USA, Canada, Saudi Arabia, UAE, Japan, Singapore, Australia and South Africa.

Regulations

The most important step towards farming taken by the government was to draw a regulatory framework. It is true that the initiatives by the government to introduce farming by laying down regulations came belatedly as many countries have already done this kind of basic work decades ago. The implementation of NPOP is ensured by the formulation of the National Accreditation Policy and Programme. The regulations make it mandatory that all organic certification bodies should be accredited by an Accreditation Agency. The international certification agencies operating in India even prior to these regulations will also have to get accreditation under the new dispensation.

The regulations lay down the institutional arrangements for implementing the national programme for organic production. The NPOP is administered, monitored and implemented for the benefit of farmers, processors, traders and consumers. It envisages a three tiered organisation under the overall guidance of the Union Government with the Department of Commerce, Ministry of Commerce and Industry as the nodal agency. Policy making and declaration of the standards for organic products, recognition of organic standards of other nations, efforts to get our standards recognized by others and coordination with other arms of the government for the successful management of the organic agriculture are the major functions entrusted to the ministry.

The agencies accredited are the Agricultural and Processed Food Products Export Development Authority, Coffee Board, Tea Board and the Spices Board. The regulations cover exports, imports and the domestic trade of the organic products. But the government regulations are applicable to only the exports. So, an organic farm product can be exported only if it is certified by a certification body accredited for the purpose. The categories of products covered under accreditation are organic crop production, organic animal production, organic processing operations, wild products and forestry. A national level steering committee is functioning as the apex advisory body for assisting the government to promote farming in the country. This body consists of representative's form the Ministries of Agriculture, Food Processing Industries, Forests and Environment, Science and Technology, Rural Development and Commerce.

Organic production requires certification after periodic inspections in order to ensure that all prescribed practices are followed. The inspection and certification are done by the agencies accredited to the Accrediting Agencies dealing with the commodity. Inspection and certification agencies can be government departments, NGOs, trade or consumer or producer organisations. Such agencies should be registered bodies, with managements in position, declaring the persons who shall be held responsible for any miscarriage of certification and having proof of adequate field staff to undertake periodic inspections. The continued accreditation of such bodies is dependent upon their record of fidelity to the principles of organic production. They are authorised to award certificates after due satisfaction that practices conformed to those enunciated by the Accrediting Agency in relation to the item concerned. The charges levied by the certifying agencies are fixed by the Accrediting Agencies.

Research and Training

A National Institute for Farming has been established to spearhead research in organic agriculture. The government of India constituted task force had also recommended the initiation of the postgraduate level courses in farming. The Morarka Foundation and Maharana Pratap University of Agriculture and Technology, Rajasthan have collaborated in the design and implementation of such a programme.

PROJECTS AND INITIATIVES

Several projects and initiatives to promote farming in the country have begun at the behest of individuals and institutions.

A project aided by the World Bank to empower the rural communities in the country to grow organic products for exports had come up in 2002. The programme aims at the improvement and promotion of organic production of spices, certification and export of black pepper, white pepper, ginger, turmeric, cardamom, clove, nutmeg and herbals like rosemary, thyme, oregano and parsley. The implementation of the progamme is done by the NGOs, and Idukki and Waynad districts of Kerala, Nilgiri district of Tamil Nadu and Kandhamal district of Orissa are the areas selected for the purpose. Imparting training to both the JNTGOs and the farmers on organic production methods, basic standards required, documentation, inspection and certification is a major objective of the programme. The assistance to NGOs includes among others computer hardware and software especially for market promotion of their produces.

An initiative for the spread of farming by various stakeholders in the Indian organic agriculture sector is their coming together to constitute an apex body for providing centralized services and expertise for the increasing number of organic farmers in the country. Initial steps for setting up the Indian Competence Centre for Organic Agriculture has been taken at a meeting held at the National Academy for Agricultural Sciences in New Delhi in 2003 under the joint auspices of INDOCERT, a Kerala based organic certification agency and the Swiss based FiBL. The decision was the outcome of an Indian team's visit to Switzerland in 2003 to study the structure of Swiss farming. The meeting elected a nine member promoter board of directors to facilitate the setting up of ICCOA, which would be registered as a charitable society. The Centre would strive to strengthen and supplement the efforts of the state governments in promoting the development of organic agriculture in the country.

In Haryana, an enterprising farmer who began farming on his 16 acre land in the Sonepat district in 1971 could establish an organic farm on 108 acres, raising vegetables and other crops. It appears that his success is mainly attributable to the efforts made by him to market the products. NGOs functioning in the neighbourhood of Delhi buy the produces like rice, wheat, pulses and vegetables from his farm at a premium price of 30 to 50 per cent. An exporting firm at a premium of 20 to 30 per cent buys his basmathi rice. Almost 70 per cent of his farm production is sold through advance agreements/ contracts. Presently he and his friends are working with the resident associations in Delhi to market their organic products at a premium of about 25 per cent.

In Rajasthan, the Morarka Foundation, established in 1995, promotes sustainable agriculture. It has about 10,000 partners producing vermi-compost and the Foundation is said to be the single largest producer of this organic input in Asia. It encourages the production of bio-pesticides and supports procurement of certified organic products. It has set up a joint venture to promote agri-biotechnology parks and a model park of such a nature has come up in Jaipur on 20 acres of land.

The campaign launched by the Foundation in the Sheldiawati region of the state in favour of farming has resulted in reduction of cultivation costs and improvement in the quality of the produces. Small and marginal farmers in the districts of Sikar and Jhunjhunu, who had been complaining of degradation of their agricultural land and declining productivity, were benefited. The efforts of the Foundation to popularize the use of vermi-compost in place of the chemical fertilizer by creating awareness and imparting training to the farmers have been successful. The Foundation too has a large facility to produce vermi-compost. Application of vermi-compost reduces irrigation, increases the flavour of the products and results in a decline in the damage to the crops by insects. The Foundation is presently engaged in developing techniques to enrich vermi-compost through micro-organisms to make it suitable for location and crop specific application.The promotion of farming is encouraged through natural resource management based on ecology protection and sustainable agricultural methods.

In Kerala, the POABS Organic Estate at Nelliyampathy is engaged in the organic cultivation of several agricultural produces. Research on farming methods including manures and pesticides is also undertaken there. Liquid manures, mixtures of slurry, cow urine and some herbal preparations are also made for use in the estate.

Several people have taken to farming in the state fully convinced of its beneficial effects on man and nature. A farmer who took farming as a mission in the district of Kottayam was an example. His estate grew only green crops, from vegetables to rubber. When the rubber plantation was raised on organic methods, the yields were low during the initial years. On the application of the organic inputs, they increased and after three years the yields were on par with the conventional rubber trees.

An initiative to make the state of Kerala fully organic has begun with the formulation of a draft policy in 2003. A workshop on Organic Sustainability of Kerala - A Global Model was organised jointly by the state agriculture department and the Confederation of Indian Industry to deliberate on the farming in Kerala. The government policy will focus on conversion of land, produces and budgetary support. Assistance will also be available for certification and inputs, promoting the local certifying bodies, development of agronomic practices, extension support and training to farmers. Creation of consumer awareness, quality considerations, and emphasis on income rather than on yield will also be part of the policy. A campaign to promote farming on the lines of the programme on Literate Kerala is proposed to be launched. Farming, it is suggested, can be promoted among the educated youth who are presently averse to conventional agriculture.

The Peerumedu Development Society is a cooperative society in the Idukki district, which promotes farming for the last 12 years. Pepper, cardamom, coffee, tea, nutmeg, clove, etc. are the crops raised by about 20000 farmers of the PDS who are encouraged to adopt farming. PDS collects the products from its member farmers and exports too. It has also started an organic tea processing factory in district.A view currently gaining ground in the state is that farming has almost become necessary for the Kerala farmer, reeling under the onslaught of the highly sensitive international markets for agriculture products. The close linking of solid waste management with bio-manure production is opening up new avenues in city planning in Kerala. The State has short listed three organizations for conducting solid waste management programmmes. The possibilities of making fuel pellets out of the rejected wastes are also being explored.

'Jaivam' or organic is the catchword now gaining popularity in Kerala. The tide seems to be turning in favour of bio-manure, bio-pest control and bio-disease control in all forms of agriculture. The state controlled Vegetable and Fruits Promotion Council, Keralam is also taking slow steps in promoting the 'bio' trend as against the overriding importance given to the chemical methods of farming.

'Infam' is an association of agriculturists functioning in the state. As a part of its efforts to promote farming, it is planning to launch a producers' company for marketing of organic produces with a view to assist the growing trend of farming in the state. In the initial stage, the functioning of the company will be confined to the Wayanad district. The company will have about 3200 farmers of the district as members who have decided to switch over to organic mode of cultivation. Steps are also being taken to forge collaboration with the Indian Institute of Rural Management, Anand. The proposed tie up with IRMA will be for technical help and also for marketing of the products. The Infam has conducted about 300 training programmes in Wayanad district, which had motivated the farmers to adopt farming methods of cultivation.

The MS Swaminathan Research Foundation, Chennai that has been operating in the area, assists the movement. Wayanad district is on the verge of an environmental calamity due to the excessive use of chemical fertilizers and pesticides that depleted the soil beyond redemption. The top soil is fast losing its fertility and is becoming barren in many areas. About 2100 banana farmers used tonnes of Furidan, an insecticide annually in the district alone. The increased use of chemicals was also posing a health hazard. A survey conducted by the Infam among the school children in Thavingal Panchayath recently revealed that 250 of them were cancer patients. The cancer disease in the region is alarmingly on the rise.

Karnataka has finalized the policy on farming and thus has become the first state in the country to adopt it. It encourages the farmers to adopt farming and gradually give up the use of chemical fertilizers and pesticides. The government has earmarked Rs 20 crores in the state budget of 2004-05 for the purpose.

The farming movement in the state has reached a stage in which the creation of markets for the products has become essential. Association for Promotion of Farming at Bangalore consists of 150 voluntary members. It has been promoting farming by creating awareness, inspecting and giving organic certification to the farms. Organic farm production in the state is not very significant. However, some NGOs and individuals are buying organic produces although the movement to convert the land to farming has not been successful. There is no organised market for organic produces and if not sold out in the right time, they are left with the conventional produce at the same prices. At the same time, while the farmers are interested in the organic cultivation the consumers are willing to pay for such products. But the link between the two simply does not exist. Consumers demand consistent supplies of good quality organic products and the farmers want an assured market for such produces.

APOP is making the farmers aware of the farming methods, particularly about the preparation of the soil for the purpose. The principles behind the healthy soil through the use of the on-farm inputs and compost making are explained to them.

Purushotthom Rao who owned 10 acres of land in the Shimoga district took up farming in 1989. He raised coconut and paddy along with several other crops in place of arecanut which was destroyed due to strong winds. The yield of coconut he obtained was 80 per cent more than that of his conventional farming friends. The cost of paddy cultivation incurred was 80 per cent lesser than his said friends.

An enterprising farmer practicing farming could export several products to the European market. His firm sent 5000 kg of black pepper, 1000 kg of nutmeg, 500 kg of mace, 4000 kg of white pepper, 1000 kg of clove, 300 kg of vanilla, 2000 kg of turmeric, 2000 kg of ginger, 5000 kg of white hibiscus, 80000 kg of henna and 500 kg of cardamom. His firm is certified by a European agency as organic since 1994.

The state of Manipur in the north-east> of India has decided to encourage organic agriculture. The north-east region with its unique characteristics and agricultural practices can be a potential area for the introduction of farming. Manipur is aiming at the encouragement of the eco-friendly farm practices and pollution free industry in the state. Dependence of the agro-chemicals will be avoided and factories and industries causing harm to plants and humans will be discouraged. The emphasis on protection of environment is in perfect harmony with the identification of the entire north-east as the mega diversity spot containing rich reserves of flora and fauna. Given the geographical location and agro-climatic zones, the state gives importance to horticulture and agro-industries. The National Horticulture Board has identified 2.77 lakh hectares for the exclusive development of horticulture in the state. It has been decided to encourage farming in this development area. The state has proposed to the setting up of an exclusive export zone for organic products in collaboration with the Agriculture Processed Food Export Development Authority.

Tamil Nadu plans to encourage farming in horticulture and plantation crops to increase the income of the farmers of the state. The state government also wants to promote organic cultivation of fruits, vegetables and tea. The Tamil Nadu Agricultural University has established a model organic farm on a 2.5 hectare area in the campus. The National Bank for Agriculture and Rural Development has decided to promote organic cultivation of horticultural crops in the mango producing areas of southern Tamil Nadu districts. Theni and Dindigul are major mango growing areas and the fact that the state government has decided to promote an agri export zone for mangoes in these districts will quicken the efforts for farming. The aim is to take the benefits of the premium prices for organic products in the developed countries through exports of mangoes.

NABARD has funded a farmer in Theni district to the tune of Rs 3.5 crores for organic mango cultivation on a 252 acre area. Identification of land for organic cultivation was a difficult process and it took one year for the farmer to do so and another three years to convince the banks. Search for the ideal planting materials took several months and he incurred an annual cost of Rs 46000 for certification.

A seminar organised at Shri Murugappa Chettiar Research Centre at Chennai focused on the need for natural cultivation of food products. Application of farm manure, biological pest control, companion planting, inter cropping, introduction of beneficial microbes and fungies were advocated for natural food farming. The importance of earthworms and the harm caused to them by chemical farming were highlighted at the discussions. In the history of farming in Tamil Nadu, the efforts made in 1970s to revive the traditional methods of agriculture at Chennai were praiseworthy. A study held by MCRC in the city revealed that 58 per cent of the people were aware of the benefits of farming and 82 per cent were ready to buy organic foods if the prices were competitive.

A few farmers in Puliangudi village in Tirunelveli district have successfully adopted a package of eco-friendly technologies in the paddy cultivation. It is found that these practices show good results on the indigenous rice varieties. The cost of cultivation has substantially been reduced and the farmers receive a premium for the organic rice. The cost of cultivation worked out comes to about Rs 8,750 per hectare. The price of rice obtained was Rs 30 per kg. The cultivation turned out to be rewarding economically besides being environmentally acceptable. Several farmers are coming forward to practice the method of farming after witnessing the results. Another farmer of the same village revealed that he got about 9,250 kg of paddy per ha and no plant protection was done after adoption of the new method. His cost of cultivation worked out was about Rs 12,500 per ha.

In Goa, a co-operative institution, the Adarsha Krishi Sahakari Kharedi Vikri Prakriya Sanstha, pioneered farming of cashew and coconut crops. The organisation has 1000 members and is led by its Chairman under whose leadership the organic cultivation is launched. The Cooperative Society has been receiving export orders from traders from abroad for supply of organic cashew nuts. The Society has helped more than 450 cashew growers to convert to farming by imparting training and other supports. It also has a processing facility for about 100 tonnes of cashew nuts during the season. The price paid for a kg of organic raw nuts is Rs 35.50. The processed nuts are taken to Mumbai and Delhi by the Society for sale.

The state of Madhya Pradesh has gone for organic soya cultivation. Plans are also ready to convert maize, wheat and pulse cultivation to organic on the principle of crop rotation. Organic soyabean is cultivated in the Malwa region of the state in the districts of Khargone and Dhar on an area of 8,700 hectares under contract farming. Farmers who have not used chemical fertilizers and pesticides during the last two years on the land offered for farming are selected. This condition was necessary to restore the soil health to make the land suitable for organic cultivation. Certification of the farm has been done by SKAL, an international agency accredited to APEDA. On getting the certificate, the farm can use the logo 'India Organic' on soyabean and its products like soya oil and deoiled cake. The production is meant for both the national and international markets. It is said that the conventional yellow soyabean fetches ₹ 12,000 per tonne in the local market, while the organic soyabean rules at ₹ 34,000 per tonne in Delhi. However, a price premium of 50 per cent is the least expected for the organic soya. This will be financially beneficial to the farmers, even though the productivity of conventional soyabean is put at 800-900 kg, which is higher than 500-600 kg of the organic soya. The genetically modified soya crops dominate the international market for soyabean and its products. However, the demand for organic soya is high ensuring a good premium for exports from the country.

PROBLEMS, CONSTRAINTS AND PROSPECTS

It is quite natural that a change in the system of agriculture in a country of more than a billion people should be a well thought out process, which requires utmost care and caution. There may be several impediments on the way. An understanding of these problems and prospects will go a long way in decision making.

PROBLEMS AND CONSTRAINTS

The most important constraint felt in the progress of farming is the inability of the government policy making level to take a firm decision to promote organic agriculture. Unless such a clear and unambiguous direction is available in terms of both financial and technical supports, from the Centre to the Panchayath levels, mere regulation making will amount to nothing. The following are found to be the major problem areas for the growth of farming in the country:

Lack of Awareness

It is a fact that many farmers in the country have only vague ideas about farming and its advantages as against the conventional farming methods. Use of bio-fertilizers and bio pesticides requires awareness and willingness on the part of the farming community. Knowledge about the availability and usefulness of supplementary nutrients to enrich the soil is also vital to increase productivity.

Farmers lack knowledge of compost making using the modern techniques and also its application. The maximum they do is making a pit and fill it with small quantities of wastes. Often the pit is flooded with rainwater and result is the top of the compost remains under composted the bottom becomes like a hard cake. Proper training to the farmers will be necessary to make vermi-compost on the modern lines.

Attention on the application of composts/organic manure is also lacking. The organic matter is spread during the months when the right moisture level is absent on the soil. The whole manure turns into wastes in the process. The required operation is of course labour intensive and costly, but it is necessary to obtain the desired results.

Output Marketing Problems

It is found that before the beginning of the cultivation of organic crops, their marketability and that too at a premium over the conventional produce has to be assured. Inability to obtain a premium price, at least during the period required to achieve the productivity levels of the conventional crop will be a setback. It was found that the farmers of organic wheat in Rajasthan got lower prices than those of the conventional wheat. The cost of marketing of both types of products was also same and the buyers of wheat were not prepared to pay higher prices to the organic variety.

Shortage of Bio-mass

Many experts and well informed farmers are not sure whether all the nutrients with the required quantities can be made available by the organic materials. Even if this problem can be surmounted, they are of the view that the available organic matter is not simply enough to meet the requirements.

The crop residues useful to prepare vermi-compost are removed after harvest from the farms.and they are used as fodder and fuel. Even if some are left out on the farms termites, etc destroy them. Experiments have shown that the crop residues ploughed back into soil will increase productivity and a better alternative is conversion into compost.

]The small and marginal cultivators have difficulties in getting the organic manures compared to the chemical fertilizers, which can be bought easily, of course if they have the financial ability. But they have to either produce the organic manures by utilizing the bio-mass they have or they have to be collected from the locality with a minimum effort and cost. Increasing pressure of population and the disappearance of the common lands including the wastes and government lands make the task difficult.

Inadequate Supporting Infrastructure

In spite of the adoption of the NPOP during 2000, the state governments are yet to formulate policies and a credible mechanism to implement them. There are only four agencies for accreditation and their expertise is limited to fruits and vegetables, tea, coffee and spices. The certiiying agencies are inadequate, the recognized green markets are non-existent, the trade channels are yet to be formed and the infrastructure facilities for verification leading to certification of the farms are inadequate.

High Input Costs

The small and marginal farmers in India have been practicing a sort of farming in the form of the traditional farming system. They use local or own farm renewable resources and carry on the agricultural practices in an ecologically friendly environment. However, now the costs of the organic inputs are higher than those of industrially produced chemical fertilizers and pesticides including other inputs used in the conventional farming system. The groundnut cake, neem seed and cake, vermi-compost, silt, cow dung, other manures, etc. applied as organic manure are increasingly becoming costly making them unaffordable to the small cultivators.

Marketing Problems of Organic Inputs

Bio-fertilizers and bio-pesticides are yet to become popular in the country. There is a lack of marketing and distribution network for them because the retailers are not interested to deal in these products, as the demand is low. The erratic supplies and the low level of awareness of the cultivators also add to the problem. Higher margins of profit for chemical fertilizers and pesticides for retailing, heavy advertisement campaigns by the manufacturers and dealers are other major problems affecting the markets for organic inputs in India.

Absence of an Appropriate Agriculture Policy

Promotion of organic agriculture both for export and domestic consumption, the requirements of food security for millions of the poor, national self-sufficiency in food production, product and input supplies, etc. are vital issues which will have to be dealt with in an appropriate agriculture policy of India. These are serious issues the solution for which hard and consistent efforts along with a national consensus will be essential to go forward. Formulation of an appropriate agriculture policy taking care of these complexities is essential to promote organic agriculture in a big way.

Lack of Financial Support

The developing countries like India have to design a plethora of national and regional standards in attune with those of the developed countries. The adoption and maintenance of such a regulatory framework and its implementation will be costly.

The cost of certification, a major component of which is the periodical inspections carried out by the certifying agencies, which have freedom to fix the timings, type and number of such inspections appears to be burdensome for the small and marginal farmers. Of course, the fees charged by the international agencies working in India before the NPOP were prohibitive and that was a reason for the weak response to organic agriculture even among the large farms in the country. No financial support as being provided in advanced countries like Germany is available in India. Supports for the marketing of the organic products are also not forthcoming neither from the State nor from the Union governments. Even the financial assistance extended to the conventional farming methods are absent for the promotion of farming.

Low Yields

In many cases the farmers experience some loss in yields on discarding synthetic inputs on conversion of their farming method from conventional to organic. Restoration of full biological activity in terms of growth of beneficial insect populations, nitrogen fixation from legumes, pest suppression and fertility problems will take some time and the reduction in the yield rates is the result in the interregnum. It may also be possible that it will take years to make organic production possible on the farm.

Small and marginal farmers cannot take the risk of low jaelds for the initial 2-3 years on the conversion to farming. There are no plans to compensate them during the gestation period. The price premiums on the organic products will not be much of help, as they will disappear once significant quantities of organic farm products are made available.

Inability to Meet the Export Demand

The demand for organic products is high in the advanced countries of the west like USA, European Union and Japan. It is reported that the US consumers are ready to pay a premium price of 60 to 100 per cent for the organic products. The upper classes in India are also following this trend as elsewhere. The market survey done by the International Trade Centre during 2000 indicates that the demand for organic products is growing rapidly in many of the world markets while the supply is unable to match it.

India is known in the world organic market as a tea supplier and there is a good potential to export coffee, vegetables, sugar, herbs, spices and vanilla. In spite of the several initiatives to produce and export organic produces from the country, the aggregate production for export came to only about 14000 tonnes.

This also includes the production of organic spices in about 1000 ha under certification. Some export houses like Good Value Marketing Ltd and Burmah Trading Corporation are also engaged in exporting of organic fruits, vegetables and coffee from India. The country could export almost 85 per cent of the production indicating that demand is not a constraint in the international markets for organic products.

Vested Interests

Hybrid seeds are designed to respond to fertilizers and chemicals. The seed, fertilizer and pesticide industry as also the importers of these inputs to the country have a stake in the conventional farming. Their opposition to farming stems from these interests.

Lack of Quality Standards for Biomanures

The need for fixing standards and quality parameters for bio-fertilizers and biomanures has arisen with the increasing popularity of farming in the country. There are a very large number of brands of organic manures, claiming the high levels of natural nutrients and essential elements. But most farmers are not aware of the pitfalls of using the commercially available biomanure products. While the concept of farming itself lays great stress on the manures produced on the farm and the farmers' household, many of the branded products available in the market may not be really organic. lements of chemicals slipping into the manures through faulty production methods could make the product not certifiable as organic. The process of composting which is a major activity to be carefully done is achieved usually by one of the two methods, vermi-composting or microbe composting.

While the former is ideal for segregated waste material without foreign matter, microbe composting is suitable for large scale management of solid wastes, especially in cities and metres. Even though the farmers are using manure produced by different methods, proper parameters for biomanure are yet to be finalized.

Most farmers are still unaware of the difference between biomanure and bio-fertilizer, it is point out. While biomanure contains organic matter, which improves the soil quality, bio-fertilizers are nutritional additives separated from the organic material, which could be added to the soil, much like taking vitamin pills. Bio-fertilizers do nothing to enhance soil quality while the loss of soil quality has been the major problem faced by farmers these days.

Improper Accounting Method

An understanding of the real costs of erosion of soil and human health, the loss of welfare of both humans and other living things and the computation of these costs are necessary to evaluate the benefits of farming. These costs will have to be integrated to a plan for the implementation of organic agriculture.

A recent study shows the inappropriateness of the cost and return accounting methods adopted to find out the economics of the farming. An economic evaluation of the bad effects of inorganic agriculture and their internalization through environmental taxes is proposed for a market based approach to promote farming in India.

Political and Social Factor

Agriculture in India is subject to political interventions with the objectives of dispensing favours for electoral benefits. Subsidies and other supports from both the Central and state governments, government controlled prices of inputs like chemical fertilizers, the public sector units' dominant role in the production of fertilizers, government support/floor prices for many agricultural products, supply of inputs like power and water either free of cost or at a subsidized rate, etc. are the tools often used to achieve political objectives. Any movement for the promotion of farming in India will have to counter opposition from the sections who benefit from such policies in the conventional farming system. The political system in a democracy like India is likely to evade the formulation of policies, which affect the interests of the voting blocks unless there are more powerful counter forces demanding changes.

In the absence of alternative employment opportunities and other considerations, the organized workforce particularly in the public sector fertilizer, pesticide and seed industries is also likely to oppose moves on the part of the government to promote farming on a large scale.

PROSPECTS

Indian agriculture should be able not only to maintain but also must strive to increase the production of foodgrains. It appears that given the availability of organic infrastructure, minimum efforts for conversion due to the low use of chemical farming methods and the limit of the public investment, farming can be progressively introduced. The potential areas and crops, which fulfill the above constraints, could be explored and brought under organic agriculture. The rainfed, tribal, north-east and hilly regions of India where the traditional farming is more or less practiced could be considered. Table gives the details of fertilizer consumption in the north-east and hilly regions of the country.

Table. Consumption of fertilizer in India in 2001-02

Sr. No.	State/Region	Quantity
1	Manipur	105
2	Jammu and Kashmir	65
3	Himachal Pradesh	41
4	Assam	39
5	Tripura	30
6	Meghalaya	17
7	Mizoram	14
8	Sikkim	10
9	Arunachal Pradesh	3

10	Nagaland	2
11	Rainfed Areas	31
	All India	90

Agriculture production in these areas is still almost on the traditional eco-friendly lines and making the farmers aware of the methods of farming may not be very difficult.

A strategy to prevent sudden and substantial yield losses is to convert to organic production in phases to reduce the risks during the initial years. The question of the vast requirement of organic matter to the country's farms in order to switch over to organic agriculture is also answered. Chemical fertilizer is applied only in 30 per cent of the cultivated area, which is irrigated, and the remaining land is under rainfed agriculture with almost no fertilizer application. Also the rainfed area under cultivation accounts for only 40 per cent of the foodgrain production of the country. The introduction of farming in these areas will allay the fears of a sudden sharp decline of food production which many fear may drive the nation to food imports. Thus the demand for biomass for the production of organic manures can also be controlled in a phased manner. Moreover, the simple technologies with low input use have been developed for dry farming and they can be transferred to the farms for organic cultivation. The resulting increases in productivity and sustain-ability of production will increasingly contribute to the betterment of the economic condition of the dry land faming community, which is one of the poorest in the country. An estimate indicates that about 600 to 700 million tonnes of biomass is available to be converted to manure. Such conversion increases the nutrient value from 0.3-04 to 1-2 per cent. Attempts can also be made to increase the supply of biomass by allocating a portion of the cultivated area to grow tree manure crops. These plants can be harvested to be used for making composts. Plans can be devised to grow green manure crops in the public lands on the lines of the social forestry programmes.

Several alternatives for supply of organic soil nutrients like vermi-composts and biofertilizers exist. Technologies have been developed to produce large quantities of these nutrients. Crop specific biofertilizers for cereals, millets, pulses and oil seeds are also available. Vermi- composting and bio-fertilizer manufacturing can be undertaken to increase the supply of organic manure to meet the demand.

The basic rules and regulations for accreditation and certification of organic products are in place in India.

A Congenial socio-cultural environment prevails in India for the promotion of organic agriculture. The farmers of India had been practicing eco-friendly agriculture for centuries till the advent of the 'green revolution' which was based on the conventional farming methods prevailed in the western countries. Still many small and marginal farmers, because of many reasons, have not fully adopted the conventional farming and they follow more or less the traditional environment friendly system. They use local or own farm derived renewable resources and manage self-regulated ecological and biological processes. This has become necessary to cultivate the acceptable levels of crop, livestock and human nutrition products and above all to protect both the crops and humans from pests and diseases through the use of bio-chemicals and bio fertilizers. Such a situation is suitable for making the farming community aware of the farming methods to make the switch over less troublesome.

A country like India can enjoy a number of benefits from the adoption of farming. The price premiums for the products, conservation of the natural resources in terms of improved soil fertility and water quality, prevention of soil erosion, preservation of natural and agro-biodiversity are major benefits. Economic and social benefits like generation of rural employment, lower urban migration, improved household nutrition, local food security and reduced dependence on external inputs will be large gains in the Indian conditions. The protection of environment and the consequent increase in the quality of human life will be other contributions of farming.

As said earlier, there is a good demand for organic products in the domestic market, which is not matched by supplies. The linkages between the two do not exist which in turn discourage production. The wholesalers/traders play a major role in the distribution of organic produces as they originate from the small farms. Large farmers have access to supermarkets and own stalls for distribution. Mumbai, Delhi, Kolkota, Chennai, Bangalore and Hyderbad are the major domestic markets for organic products. High premium prices both in the national and international markets and the lack of supplies in the latter reveal the opportunities for India. Similarly the case of organic cotton of India is judged as having bright prospects.

Organic cotton/eco-cotton/green cotton is the cotton grown without inorganic fertilizers, pesticides and defoliants and duly certified by a recognized certifying organization. India has tremendous potential to emerge as a world leader in organic cotton. Indian cotton textile industry is one of the oldest organized enterprises and is supported by supply of raw cotton cultivated in the country. Cotton based products come to about 35 per cent of the country's exports. The present level of cotton productivity is only about 300 kg lint/ha against a world average of 600 kg. The low productivity is attributed to a large area under rainfed farming - almost 65 per cent of the cotton area.

India' has the largest area under cotton in the world, but it is only the third in terms of production. The productivity is low even after 75 per cent of the cotton area is covered by the high yielding varieties. The reason for low productivity among others is the diseases/pest attacks on the crop as a result of the conversion of the traditional varieties to the high yielding ones. India needs to increase cotton production to meet both the domestic and international demand. This has to come from an increase in productivity, which presently is very low. India is the only producer of all the four species of cotton in the world. Not only this, the country produces the widest quality range of cotton suitable for spinning 6's to 12's count yarn which can meet the demand from both the domestic and international textile industries. During the first half of the last century, the country grew only the desi cotton and the American variety accounted for a small area only. The desi variety yields are low, but they require lesser moisture, resist pests and diseases to a considerable extent, and are suita.ble in dry lands. And this cotton was organically cultivated. The increasing demand for cotton pushed out the desi cotton from cultivation and was replaced by the high yielding varieties. The desi cotton is grown only on 25 per cent of the total area under cotton at present.

The conventional method of cultivation of cotton needs intensive production technologies requiring very high levels of inputs like fertilizers, insecticides, and water for irrigation. The inputs are often excessively used causing ecological problems besides diminishing rate of returns.

Use of pesticides in the cotton crop is very intense and several chemicals are used indiscriminately ignoring the environment. Even the chemicals, banned/restricted for use in the advanced countries, are used in the Indian cotton crop.

Organic cultivation of cotton under certification will be profitable as there is a strong demand for eco-cotton in many Western and Asian countries. There are about 12 countries producing eco-cotton. USA, Greece, Israel, Peru, Egypt, Turkey, China and Australia are important eco-cotton producing countries. Eco-cotton commands a price higher than 30 to 40 per cent of the conventional cotton. However, the cotton growers in the country will take up organic cultivation only if the yield loss if any as a result of the switch over is compensated by high prices and the technology is feasible and financially viable.

Organic cotton can be produced in India easily in those areas where cotton is raised under the dry farming conditions. These areas are under the desi variety and as such can be converted into organic cotton farming easily. But the potential areas are spread over several states of western and southern India. Coloured cotton was cultivated in Andhra Pradesh prior to 1950. High quality hand woven fabrics were made from desi cotton once upon the time. But white cotton has replaced the coloured ones during the last fifty years. Now there is a demand world over for organically produced coloured cotton and India is in a position to cater to the international market.

CEASES

The ill effects of the conventional farming system are felt in India in terms of the unsustainablity of agricultural production, environmental degradation, health and sanitation problems, etc. Organic agriculture is gaining momentum as an alternative method to the modern system. Many countries have been able to convert 2-10 per cent of their cultivated areas into farming. The demand for organic products is growing fast.

It appears that India is lagging far behind in the adoption of farming. So far, the only achievement seems to be the laying down of the National Standards for Organic Production and the approval of 4 accreditation agencies whose expertise is limited to a few crops.

The following aresome of the issues, which require attention at the government policy making levels if we want to lay the spadework for the spread of organic agriculture in the country.

- Substantial financial support by governments is absolutely necessary to promote farming. A major factor behind the progress made by the major organic countries has been the very liberal subsidies provided by the governments. In India, organic farmers do not receive the benefits of government subsidies as they are targeted at the conventional cultivation. Given the low risk bearing capacity, the need to make the farming an attractive proposition at least during the initial period, the likely prospect of loss of productivity for some time, and the non-existence of marketing channels for organic produces the financial support must be adequate. The suggestion to begin the introduction of farming in the north-east region and in the dry farming areas of India is a well thought out proposition. However, it should be remembered that these regions are inhabited by the poorest and least advantaged groups whose dependence on agriculture for a livelihood is total. A programme for organic agriculture in these areas must be fully supported by the full compensation both in cash and kind to the farmers in the event of the loss of production they would suffer till it reaches the levels of the days prior to the adoption of farming. A fair, quick and efficient delivery system for such assistance, perhaps by keeping the government bureaucracy at a distance should also be in place beforehand.
- Market development for the organic products is a crucial factor to promote domestic sales. Supplies do not match the demand for organic products in the country and the absence of proper links between the two has been pointed out for the tardy growth of farming in the country. An important role of the government in this direction is giving various supports to the producer and consumer associations to market the products.
- The producer organizations must be encouraged to get accredited for inspection and certification in accordance with the NSOP. They can also have own standards and even symbols. This may also reduce the costs of certification besides the simplification of the process.

- A vigorous campaign to highlight the benefits of farming against the conventional system is essential to increase the awareness of the farmers and consumers.
- Identification of crops for cultivation on the organic farms is important. The examples of soyabean in Madhya Pradesh and cotton in the rainfed areas could be kept in view in the process.

5

Bank Supports for Diary Faming

Dairying is an important source of subsidiary income to small/marginal farmers and agricultural labourers. The manure from animals provides a good source of organic matter for improving soil fertility and crop yields. The gober gas from the dung is used as fuel for domestic purposes as also for running engines for drawing water from well. The surplus fodder and agricultural by-products are gainfully utilised for feeding the animals.

Almost all draught power for farm operations and transportation is supplied by bullocks. Since agriculture is mostly seasonal, there is a possibility of finding employment throughout the year for many persons through dairy farming.

Thus, dairy also provides employment throughout the year. The main beneficiaries of dairy programmes are small/marginal farmers and landless labourers. A farmer can earn a gross surplus of about ₹ 72,000 per year from a unit consisting of 2 milking buffaloes.

The capital investment required for purchase of 2 buffaloes is ₹ 108,223/-. Even after paying a sum of ₹ 36c294/- per annum towards repayment of the loan and interest the farmer can earn a net surplus of ₹ 72,000 - 108,000/- approximately per year. Even more profits can be earned depending upon the breed of animal, managerial skills and marketing potential.According to World Bank estimates about 75 per cent of India's 940 million people are in 5.87 million villages, cultivating over 145 million hectares of cropland. Average farm size is about 1.66 hectares. Among 70 million rural households, 42 per cent operate upto 2 hectares and 37 per cent are landless households.

These landless and small farmers have in their possession 53 per cent of the animals and produce 51 per cent of the milk. Thus, small/marginal farmers and land less agricultural labourers play a very important role in milk production of the country. Dairy farming can also be taken up as a main occupation around big urban centres where the demand for milk is high.

SCOPE FOR DAIRY FARMING AND ITS NATIONAL IMPORTANCE

The total milk production in the country for the year 2001-02 was estimated at 84.6 million metric tonnes. At this production, the per capita availability was to be 226 grams per day against the minimum requirement of 250 grams per day as recommended by ICMR. Thus, there is a tremendous scope/potential for increasing the milk production. The population of breeding cows and buffaloes in milk over 3 years of age was 62.6 million and 42.4 million, respectively (1992 census)

Central and State Governments are giving considerable financial assistance for creating infrastructure facilities for milk production. The nineth plan outlay on Animal Husbandry and Dairying was ₹ 2345 crores.

FINANCIAL ASSISTANCE AVAILABLE FROM BANKS/NABARD FOR DAIRY FARMING

NABARD is an apex institution for all matters relating to policy, planning and operation in the field of agricultural credit. It serves as an apex refinancing agency for the institutions providing investment and production credit. It promotes development through formulation and appraisal of projects through a well organised Technical Services Department at the Head Office and Technical Cells at each of the Regional Offices. Loan from banks with refinance facility from NABARD is available for starting dairy farming. For obtaining bank loan, the farmers should apply to the nearest branch of a commercial or co-operative Bank in their area in the prescribed application form which is available in the branches of financing banks. The Technical Officer attached to or the Manager of the bank can help/give guidance to the farmers in preparing the project report to obtain bank loan.

For dairy schemes with very large outlays, detailed reports will have to be prepared. The items of finance would include capital asset items such as purchase of milch animals, construction of sheds, purchase of equipments etc. The feeding cost during the initial period of one/two months is capitalised and given as term loan.

Facilities such as cost of land development, fencing, digging of well, commissioning of diesel engine/pumpset, electricity connections, essential servants' quarters, godown, transport vehicle, milk processing facilities etc. can be considered for loan. Cost of land is not considered for loan. However, if land is purchased for setting up a dairy farm, its cost can be treated as party's margin upto 10% of the total cost of project.

SCHEME FORMULATION FOR BANK LOAN

A Scheme can be prepared by a beneficiary after consulting local technical persons of State animal husbandry department, DRDA, SLPP etc., dairy co-operative society/union/federation/commercial dairy farmers.If possible, the beneficiaries should also visit progressive dairy farmers and government/

military/agricultural university dairy farm in the vicinity and discuss the profitability of dairy farming. A good practical training and experience in dairy farming will be highly desirable. The dairy co-operative societies established in the villages as a result of efforts by the Dairy Development Department of State Government and National Dairy Development Board would provide all supporting facilities particularly marketing of fluid milk. Nearness of dairy farm to such a society, veterinary aid centre, artificial insemination centre should be ensured. There is a good demand for milk, if the dairy farm is located near urban centre.The scheme should include information on land, livestock markets, availability of water, feeds, fodders, veterinary aid, breeding facilities, marketing aspects, training facilities, experience of the farmer and the type of assistance available from State Government, dairy society/union/federation.

The scheme should also include information on the number of and types of animals to be purchased, their breeds, production performance, cost and other relevant input and output costs with their description. Based on this, the total cost of the project, margin money to be provided by the beneficiary, requirement of bank loan, estimated annual expenditure, income, profit and loss statement, repayment period, etc..

SCRUTINY OF SCHEMES BY BANKS

The scheme so formulated should be submitted to the nearest branch of bank. The bank's officers can assist in preparation of the scheme for filling in the prescribed application form. The bank will then examine the scheme for its technical feasibility and economic viability.

TECHNICAL FEASIBILITY

- Nearness of the selected area to veterinary, breeding and milk collection centre and the financing bank's branch.
- Availability of good quality animals in nearby livestock market.
- Availability of training facilities.
- Availability of good grazing ground/lands.
- Green/dry fodder, concentrate feed, medicines etc.
- Availability of veterinary aid/breeding centres and milk marketing facilities near the scheme area.

ECONOMIC VIABILITY

- Unit Cost.
- Input cost for feeds and fodders, veterinary aid, breeding of animals, insurance, labour and other overheads.
- Output costs *i.e.* sale price of milk, manure, gunny bags, male/female calves, other miscellaneous items etc.
- Income-expenditure statement and annual gross surplus.

- Cash flow analysis.
- Repayment schedule (*i.e.* repayment of principal loan amount and interest).

Other documents such as loan application forms, security aspects, margin money requirements etc. are also examined. A field visit to the scheme area is undertaken for conducting a techno-economic feasibility study for appraisal of the scheme.

SANCTION OF BANK LOAN AND ITS DISBURSEMENT

After ensuring technical feasibility and economic viability, the scheme is sanctioned by the bank. The loan is disbursed in kind in 2 to 3 stages against creation of specific assets such as construction of sheds, purchase of equipments and machinery, purchase of animals and recurring cost on purchase of feeds/ fodders for the initial period of one/two months. The end use of the fund is verified and constant follow-up is done by the bank.

LENDING TERMS – GENERAL

UNIT COST

Each Regional Office (RO) of NABARD has constituted a State Level Unit Cost Committee under the Chairmanship of RO-in-charges and with the members from developmental agencies, commercial banks and cooperative banks to review the unit cost of various investments once in six months. The same is circulated among the banks for their guidance. These costs are only indicative in nature and banks are free to finance any amount depending upon the availability of assets.

MARGIN MONEY

NABARD had defined farmers into three different categories and where subsidy is not available the minimum down payment as shown below is collected from the beneficiaries.

Sr.No Beneficiary's	Category of Farmer return to resources	Level of predevelopment Contribution
(a)	Small Farmers	Upto Rs.11000 5%
(b)	Medium Farmers	Rs.11001 - Rs.19250 10%
(c)	Large Farmers	Above ₹ 19251 15%'

INTEREST RATE

As per the RBI guidelines the present rate of interest to the ultimate beneficiary financed by various agencies are as under:

No.	Loan Amount	CB's and RRB's	SLDB/SCB
(a)	Upto and inclusive	12%	As determined by

	of Rs.25000 to minimum 12%	12%	SCB/SLDB subject
(b)	Over ₹ 25000 and upto ₹ 2 lakhs	13.5% 13.5%	-do-
(c)	Over ₹ 2.0 lakhs	As determined by the banks	-do-

SECURITY

Security will be as per NABARD/RBI guidelines issued from time to time.

REPAYMENT PERIOD OF LOAN

Repayment period depends upon the gross surplus in the scheme. The loans will be repaid in suitable monthly/quarterly instalments usually within a period of about 5 years. In case of commercial schemes it may be extended upto 6-7 years depending on cash flow analysis.

INSURANCE

The animals may be insured annually or on long term master policy, where ever it is applicable. The present rate of insurance premium for scheme and Non- scheme animals are 2.25% and 4.0% respectively.

PACKAGE OF COMMON MANAGEMENT PRACTICES RECOMMENDED FOR DAIRY

FARMERS

Modern and well established scientific principles, practices and skills should be used to obtain maximum economic benefits from dairy farming. Some of the major norms and recommended practices are as follows:

I Housing

- Construct shed on dry, properly raised ground.
- Avoid water-logging, marshy and heavy rainfall areas.
- The walls of the sheds should be 1.5 to 2 metres high.
- The walls should be plastered to make them damp proof.
- The roof should be 3-4 metres high.
- The cattle shed should be well ventilated.
- The floor should be pucca/hard, even non-slippery impervious, well sloped (3 cm per metre) and properly drained to remain dry and clean.
- Provide 0.25 metre broad, pucca drain at the rear of the standing space.
- A standing space of 2 x 1.05 metre for each animal is needed.
- The manger space should be 1.05 metre with front height of 0.5 metre and depth of 0.25 metre.

- The corners in mangers, troughs, drains and walls should be rounded for easy cleaning.
- Provide 5-10 sq. metre loaf space for each animal.
- Provide proper shade and cool drinking water in summer.
- In winter keep animals indoor during night and rain.
- Provide individual bedding daily.
- Maintain sanitary condition around shed.
- Control external parasites (ticks, flies etc.) by spraying the pens, sheds with Malathion or Copper sulphate solution.
- Drain urine into collection pits and then to the field through irrigation channels.
- Dispose of dung and urine properly. A gobar gas plant will be an ideal way. Where gobar gas plant is not constructed, convert the dung alongwith bedding material and other farm wastes into compost.
- Give adequate space for the animals.

II. Selection of Animal

- Immediately after release of the loan purchase the stock from a reliable breeder or from nearest livestock market.
- Select healthy, high yielding animals with the help of bank's technical officer, veterinary/animal husbandry officer of State government/Zilla Parishad, etc.
- Purchase freshly calved animals in their second/third lactation.
- Before purchasing, ascertain actual milk yield by milking the animal three times consecutively.
- Identify the newly purchased animal by giving suitable identification mark (ear tagging or tattooing).
- Vaccinate the newly purchased animal against disease.
- Keep the newly purchased animal under observation for a period of about two weeks and then mix with the general herd.
- Purchase a minimum economical unit of two milch animals.
- Purchase the second animal/second batch after 5-6 months from the purchase of first animal.
- As buffaloes are seasonal calvers purchase them during July to February.
- As far as possible purchase the second animal when the first animal is in its late stage of lactation and is about to become dry, thereby maintaining continuity in milk production vis-a-vis income. This will ensure availability of adequate funds for maintaining the dry animals.
- Follow judicious culling and replacement of animals in a herd.
- Cull the old animals after 6-7 lactations.

III. Feeding of Milch Animals

- Feed the animals with best feeds and fodders. (Feeding schedule is given in Anneuxre VIII).
- Give adequate green fodder in the ration.
- As far as possible, grow green fodder on your land wherever available.
- Cut the fodder at the right stage of their growth.
- Chaff roughage before feeding.
- Crush the grains and concentrates.
- The oil cakes should be flaky and crumbly.
- Moisten the concentrate mixture before feeding.
- Provide adequate vitamins and minerals. Provide salt licks besides addition of mineral mixture to the concentrate ration.
- Provide adequate and clean water.
- Give adequate exercise to the animals. Buffaloes should be taken for wallowing daily. In case this is not possible sprinkle sufficient water more particularly during summer months.
- To estimate the daily feed requirement remember that the animals consume about 2.5 to 3.0 per cent of their body weight on dry matter basis.

IV. Milking of Animals

- Milk the animals two to three times a day.
- Milk at fixed times.
- Milk in one sitting within eight minutes.
- As far as possible, milking should be done by the same person regularly.
- Milk the animal in a clean place.
- Wash the udder and teat with antiseptic lotions/luke-warm water and dry before milking.
- Milker should be free from any contagious diseases and should wash his hands with antiseptic lotion before each milking.
- Milking should be done with full hands, quickly and completely followed by stripping.
- Sick cows/buffaloes should be milked at the end to prevent spread of infection.

V. Protection against Diseases

- Be on the alert for signs of illness such as reduced feed intake, fever, abnormal discharge or unusual behaviour.
- Consult the nearest veterinary aid centre for help if illness is suspected.
- Protect the animals against common diseases.
- In case of outbreak of contagious disease, immediately segregate the sick, in-contact and the healthy animals and take necessary disease control measures.
- Conduct periodic tests for Brucellosis, Tuberculosis, Johne's disease, Mastitis etc.

- Deworm the animals regularly.
- Examine the faeces of adult animals to detect eggs of internal parasites and treat the animals with suitable drugs.
- Wash the animals from time to time to promote sanitation.

VI. Breeding Care

- Observe the animal closely and keep specific record of its coming in heat, duration of heat, insemination, conception and calving.
- Breed the animals in time.
- The onset of oestrus will be within 60 to 80 days after calving.
- Timely breeding will help achieving conception within 2 to 3 months of calving.
- Breed the animals when it is in peak heat period (*i.e.* 12 to 24 hours of heat).
- Use high quality semen preferably frozen semen of proven sires/bulls.

VII. Care during Pregnancy

Give special attention to pregnant cows two months before calving by providing adequate space, feed, water etc.

VIII. Marketing of Milk

- Marketing milk immediately after it is drawn keeping the time between production and marketing of the milk to the minimum.
- Use clean utensils and handle milk in hygienic way.
- Wash milk pails/cans/utensils thoroughly with detergent and finally rinse with chloride solution.
- Avoid too much agitation of milk during transit.
- Transport the milk during cool hours of the day.

IX. Care of Calves

- Take care of new born calf.
- Treat/disinfect the navel cord with tincutre of iodine as soon as it is cut with a sharp knife.
- Feed colostrum to calf.
- Assist the calf to suckle if it is too weak to suckle on its own within 30 minutes of calving.
- In case it is desired to wean the calf immediately after birth, then feed the colostrum in bucket.
- Keep the calf separately from birth till two months of age in a dry clean and well ventilated place.
- Protect the calves against extreme weather conditions, particularly during the first two months.
- Group the calves according to their size.

- Vaccinate calves.
- Dehorn the calves around 4 to 5 days of age for easy management when they grow.
- Dispose of extra calves not to be reared/maintained for any specific purpose as early as possible, particularly the male calves.
- The female calves should be properly reared.

FORMAT FOR SUBMISSION OF SCHEMES

GENERAL

- Name of the sponsoring bank
- Address of the controlling ofice sponsoring the scheme
- Nature and objectives of the proposed scheme
- Details of proposed investments

S.No	Investment	No. Of units
(a)		
(b)		
(c)		

- Specification of the scheme area (Name of District and Block/s)

S.No.	District	Block

- Names of the financing bank's branches

S.No.	Name of the Branch/District
(a)	
(b)	
(c)	

- *Status of beneficiary/ies*: (Indidivual/Partnership/Company/Corporation/ Co-operative Society/Others)
- In case of area based schemes, coverage of borrowers in weaker parts (landless labourers, small, medium and large farmers as per NABARD's norms, SC/ST, etc.)
- Details of borrowers profile (Not applicable to area based schemes)
  - Capability
  - Experience
  - Financial Soundness
  - Technical/Other special Qualificaitons
  - Technical/Managerial Staff and adequacy thereof

TECHNICAL ASPECTS

a) *Location, Land and Land Development*:
   - Location details of the project
   - Total Area of land and its cost

- Site map
- Particulars of land development, fencing, gates, etc.

b) *Civil Structures*:
- Sheds
- Store room
- Milk room
- Quarters, etc.

c) *Equipment/Plant and Machinery*:
- Chaff cutter
- Silo pit
- Milking machine
- Feed grinder and mixer
- Milking pails/milk cans
- Biogas plant
- Bulk coolers
- viii)Equipment for manufacture of products
- Truck/van (price quotations for the above equipments)

d) *Housing*:
- Type of housing
- Area requirement
- Adults
- Heifers (1-3 years)
- Calves (less than 1 year)

e) *Animals:*
- Proposed species
- Proposed breed
- Source of purchase
- Place of purchase
- Distance (kms.)
- Cost of animal (₹)

f) *Production parametres*:
- Order of lactation
- Milk yield (ltrs. per day)
- Lactation days
- Dry days
- Conception rate
- Mortality(per cent)
- Adults
- Young stock

g) *Herd projection (with all assumptions)*:

h) *Feeding*:

- Source of fodder and feed - Green fodder
- – Dry fodder
- – Concentrates
- Fodder crop rotations
- – Kharif
- – Rabi
- – Summer

- Fodder cultivation expenses
- Requirement and costs:

Quantity Required (kg./day)

	Cost(Rs. / Kg)	Lactation	Dry Period	Young Stock
Green Fodder				
Dry Fodder				
Concentrates				

i) *Breeding Facilities*:

- Source
- Location
- Distance (km.)
- Availability of semen
- Availability of staff
- Expenditure per animal/year

j) *Veterinary Aid*:

- Source
- Location
- Distance (km.)
- Availability of staff
- Types of facilities available
- If own arrangements are made -
- – Employed a veterinary doctor/stockman/consultant
- – Periodicity of visit
- – Amount paid/visit (₹)
- Expenditure per animal per year (₹)

k) *Electricity*:

- Source
- Approval from SEB
- Connected load
- Problems of power failure
- Arrangements for generator

l) *Water*:

- Source
- Quality of water
- Abvailability of sufficient quantity for drinking, cleaning nad fodder production
- If investment has to be made, type of strucutre, design and cost

m) *Marketing of milk*:
- Source of sales
- Place of disposal
- Distance (km.)
- Price realised - (₹ per litre of milk)
- Basis of payment
- Periodicity of paymen

n) *Marketing of other products*:
- *Animal*: Age
- – Place of sale
- – Price expected
- *Manure*: Qty./animal
  - – Price/unit (₹)
- Empty gunny bags
- – Number
- – Cost/bag (₹)

o) *Beneficiary's experience*:

p) *Comments on technical feasibility*:

q) *Government restrictions, if any*:

6

Uses of Dairy Farming for Small Farmer

Dairying is an important source of subsidiary income to small/marginal farmers and agricultural labourers. The manure from animals provides a good source of organic matter for improving soil fertility and crop yields. The gober gas from the dung is used as fuel for domestic purposes as also for running engines for drawing water from well. The surplus fodder and agricultural by-products are gainfully utilised for feeding the animals.

Almost all draught power for farm operations and transportation is supplied by bullocks. Since agriculture is mostly seasonal, there is a possibility of finding employment throughout the year for many persons through dairy farming. Thus, dairy also provides employment throughout the year. The main beneficiaries of dairy programmes are small/marginal farmers and landless labourers. A farmer can earn a gross surplus of about ₹ 12,000 per year from a unit consisting of 2 milking buffaloes.

The capital investment required for purchase of 2 buffaloes is ₹ 18,223/-. Even after paying a sum of ₹ 4294/- per annum towards repayment of the loan and interest the farmer can earn a net surplus of ₹ 6000 - 9000/- approximately per year. Even more profits can be earned depending upon the breed of animal, managerial skills and marketing potential. World Bank estimates about 75 per cent of India's 940 million people are in 5.87 million villages, cultivating over 145 million hectares of cropland. Average farm size is about 1.66 hectares. Among 70 million rural households, 42 per cent operate upto 2 hectares and 37 per cent are landless households.

These landless and small farmers have in their possession 53 per cent of the animals and produce 51 per cent of the milk. Thus, small/marginal farmers and land less agricultural labourers play a very important role in milk production of the country. Dairy farming can also be taken up as a main occupation around big urban centres where the demand for milk is high.

SCOPE FOR DAIRY FARMING AND ITS NATIONAL IMPORTANCE

The total milk production in the country for the year 2001-02 was estimated at 84.6 million metric tonnes. At this production, the per capita availability was to be 226 grams per day against the minimum requirement of 250 grams per day as recommended by ICMR. Thus, there is a tremendous scope/potential for increasing the milk production. The population of breeding cows and buffaloes in milk over 3 years of age was 62.6 million and 42.4 million, respectively (1992 census). Central and State Governments are giving considerable financial assistance for creating infrastructure facilities for milk production. The nineth plan outlay on Animal Husbandry and Dairying was ₹ 2345 crores.

FINANCIAL ASSISTANCE AVAILABLE FROM BANKS/NABARD FOR DAIRY FARMING

NABARD is an apex institution for all matters relating to policy, planning and operation in the field of agricultural credit. It serves as an apex refinancing agency for the institutions providing investment and production credit. It promotes development through formulation and appraisal of projects through a well organised Technical Services Department at the Head Office and Technical Cells at each of the Regional Offices.

Loan from banks with refinance facility from NABARD is available for starting dairy farming. For obtaining bank loan, the farmers should apply to the nearest branch of a commercial or co-operative Bank in their area in the prescribed application form which is available in the branches of financing banks. The Technical Officer attached to or the Manager of the bank can help/give guidance to the farmers in preparing the project report to obtain bank loan.

For dairy schemes with very large outlays, detailed reports will have to be prepared. The items of finance would include capital asset items such as purchase of milch animals, construction of sheds, purchase of equipments etc. The feeding cost during the initial period of one/two months is capitalised and given as term loan. Facilities such as cost of land development, fencing, digging of well, commissioning of diesel engine/pumpset, electricity connections, essential servants' quarters, godown, transport vehicle, milk processing facilities etc. can be considered for loan. Cost of land is not considered for loan. However, if land is purchased for setting up a dairy farm, its cost can be treated as party's margin upto 10% of the total cost of project.

SCHEME FORMULATION FOR BANK LOAN

A Scheme can be prepared by a beneficiary after consulting local technical persons of State animal husbandry department, DRDA, SLPP etc., dairy co-operative society/union/federation/commercial dairy farmers. If possible, the beneficiaries should also visit progressive dairy farmers and government/ military/agricultural university dairy farm in the vicinity and discuss the profitability of dairy farming. A good practical training and experience in dairy farming will be highly desirable. The dairy co-operative societies established in the villages as a result of efforts by the Dairy Development Department of State Government and National Dairy Development Board would provide all supporting facilities particularly marketing of fluid milk. Nearness of dairy farm to such a society, veterinary aid centre, artificial insemination centre should be ensured. There is a good demand for milk, if the dairy farm is located near urban centre.

The scheme should include information on land, livestock markets, availability of water, feeds, fodders, veterinary aid, breeding facilities, marketing aspects, training facilities, experience of the farmer and the type of assistance available from State Government, dairy society/union/federation.

The scheme should also include information on the number of and types of animals to be purchased, their breeds, production performance, cost and other relevant input and output costs with their description. Based on this, the total cost of the project, margin money to be provided by the beneficiary, requirement of bank loan, estimated annual expenditure, income, profit and loss statement, repayment period, etc. can be worked out and shown in the Project report. A format developed for formulation of dairy development schemes.

SCRUTINY OF SCHEMES BY BANKS

The scheme so formulated should be submitted to the nearest branch of bank. The bank's officers can assist in preparation of the scheme for filling in the prescribed application form. The bank will then examine the scheme for its technical feasibility and economic viability.

Technical Feasibility—this would briefly include:

- Nearness of the selected area to veterinary, breeding and milk collection centre and the financing bank's branch.
- Availability of good quality animals in nearby livestock market. The distribution of important breeds of cattle and buffaloes. The reproductive and productive performance of cattle and buffalo breeds.
- Availability of training facilities.
- Availability of good grazing ground/lands.
- Green/dry fodder, concentrate feed, medicines etc.
- Availability of veterinary aid/breeding centres and milk marketing facilities near the scheme area.

Economic Viability—this would briefly include:

- *Unit Cost*: The average unit cost of dairy animals for some of the States.
- Input cost for feeds and fodders, veterinary aid, breeding of animals, insurance, labour and other overheads.

- Output costs *i.e.* sale price of milk, manure, gunny bags, male/female calves, other miscellaneous items etc.
- Income-expenditure statement and annual gross surplus.
- Cash flow analysis.
- Repayment schedule (*i.e.* repayment of principal loan amount and interest).

Other documents such as loan application forms, security aspects, margin money requirements etc. are also examined. A field visit to the scheme area is undertaken for conducting a techno-economic feasibility study for appraisal of the scheme. Model economics for a two animal unit and mini dairy unit with ten buffaloes.

SANCTION OF BANK LOAN AND ITS DISBURSEMENT

After ensuring technical feasibility and economic viability, the scheme is sanctioned by the bank. The loan is disbursed in kind in 2 to 3 stages against creation of specific assets such as construction of sheds, purchase of equipments and machinery, purchase of animals and recurring cost on purchase of feeds/ fodders for the initial period of one/two months.

The end use of the fund is verified and constant follow-up is done by the bank.

LENDING TERMS—GENERAL

UNIT COST

Each Regional Office (RO) of NABARD has constituted a State Level Unit Cost Committee under the Chairmanship of RO-in-charges and with the members from developmental agencies, commercial banks and cooperative banks to review the unit cost of various investments once in six months.

The same is circulated among the banks for their guidance. These costs are only indicative in nature and banks are free to finance any amount depending upon the availability of assets.

MARGIN MONEY

NABARD had defined farmers into three different categories and where subsidy is not available the minimum down payment as shown below is collected from the beneficiaries.

Sr.No.	Category of Farmer	Level of predevelopment return to resources	Beneficiary's Contribution
(a)	Small Farmers	Upto Rs.11000	5%
(b)	Medium Farmers	Rs.11001 - Rs.19250	10%
(c)	Large Farmers	Above ₹ 19251	15%`

INTEREST RATE

As per the RBI guidelines the present rate of interest to the ultimate beneficiary financed by various agencies are as under:

No.	Loan Amount	CB's and RRB's	SLDB/SCB
(a)	Upto and inclusive of Rs.25000 to minimum 12%	12%	As determined by SCB/SLDB subject
(b)	Over ₹ 25000 and upto ₹ 2 lakhs	13.5%	-do-
(c)	Over ₹ 2.0 lakhs	As determined by the banks	-do-

SECURITY

Security will be as per NABARD/RBI guidelines issued from time to time.

REPAYMENT PERIOD OF LOAN

Repayment period depends upon the gross surplus in the scheme. The loans will be repaid in suitable monthly/quarterly instalments usually within a period of about 5 years. In case of commercial schemes it may be extended upto 6-7 years depending on cash flow analysis.

INSURANCE

The animals may be insured annually or on long term master policy, where ever it is applicable. The present rate of insurance premium for scheme and Non- scheme animals are 2.25% and 4.0% respectively.

PACKAGE OF COMMON MANAGEMENT PRACTICES RECOMMENDED FOR DAIRY

FARMERS

Modern and well established scientific principles, practices and skills should be used to obtain maximum economic benefits from dairy farming.

Some of the major norms and recommended practices are as follows:

I. Housing

- Construct shed on dry, properly raised ground.
- Avoid water-logging, marshy and heavy rainfall areas.
- The walls of the sheds should be 1.5 to 2 metres high.
- The walls should be plastered to make them damp proof.
- The roof should be 3-4 metres high.
- The cattle shed should be well ventilated.
- The floor should be pucca/hard, even non-slippery impervious, well sloped (3 cm per metre) and properly drained to remain dry and clean.

- Provide 0.25 metre broad, pucca drain at the rear of the standing space.
- A standing space of 2 × 1.05 metre for each animal is needed.
- The manger space should be 1.05 metre with front height of 0.5 metre and depth of 0.25 metre.
- The corners in mangers, troughs, drains and walls should be rounded for easy cleaning.
- Provide 5-10 sq. metre loaf space for each animal.
- Provide proper shade and cool drinking water in summer.
- In winter keep animals indoor during night and rain.
- Provide individual bedding daily.
- Maintain sanitary condition around shed.
- Control external parasites (ticks, flies etc.) by spraying the pens, sheds with Malathion or Copper sulphate solution.
- Drain urine into collection pits and then to the field through irrigation channels.
- Dispose of dung and urine properly. A gobar gas plant will be an ideal way. Where gobar gas plant is not constructed, convert the dung alongwith bedding material and other farm wastes into compost.
- Give adequate space for the animals. .

II. Selection of Animal

- Immediately after release of the loan purchase the stock from a reliable breeder or from nearest livestock market.
- Select healthy, high yielding animals with the help of bank's technical officer, veterinary/animal husbandry officer of State government/Zilla Parishad, etc.
- Purchase freshly calved animals in their second/third lactation.
- Before purchasing, ascertain actual milk yield by milking the animal three times consecutively.
- Identify the newly purchased animal by giving suitable identification mark (ear tagging or tattooing).
- Vaccinate the newly purchased animal against disease.
- Keep the newly purchased animal under observation for a period of about two weeks and then mix with the general herd.
- Purchase a minimum economical unit of two milch animals.
- Purchase the second animal/second batch after 5-6 months from the purchase of first animal.
- As buffaloes are seasonal calvers purchase them during July to February.
- As far as possible purchase the second animal when the first animal

is in its late stage of lactation and is about to become dry, thereby maintaining continuity in milk production vis-a-vis income. This will ensure availability of adequate funds for maintaining the dry animals.

- Follow judicious culling and replacement of animals in a herd.
- Cull the old animals after 6-7 lactations.

III. Feeding of Milch Animals

- Feed the animals with best feeds and fodders. (Feeding schedule is given in Anneuxre VIII).
- Give adequate green fodder in the ration.
- As far as possible, grow green fodder on your land wherever available.
- Cut the fodder at the right stage of their growth.
- Chaff roughage before feeding.
- Crush the grains and concentrates.
- The oil cakes should be flaky and crumbly.
- Moisten the concentrate mixture before feeding.
- Provide adequate vitamins and minerals. Provide salt licks besides addition of mineral mixture to the concentrate ration.
- Provide adequate and clean water.
- Give adequate exercise to the animals. Buffaloes should be taken for wallowing daily. In case this is not possible sprinkle sufficient water more particularly during summer months.
- To estimate the daily feed requirement remember that the animals consume about 2.5 to 3.0 per cent of their body weight on dry matter basis.

IV. Milking of Animals:

- Milk the animals two to three times a day.
- Milk at fixed times.
- Milk in one sitting within eight minutes.
- As far as possible, milking should be done by the same person regularly.
- Milk the animal in a clean place.
- Wash the udder and teat with antiseptic lotions/luke-warm water and dry before milking.
- Milker should be free from any contagious diseases and should wash his hands with antiseptic lotion before each milking.
- Milking should be done with full hands, quickly and completely followed by stripping.
- Sick cows/buffaloes should be milked at the end to prevent spread of infection.

V. Protection against Diseases

- Be on the alert for signs of illness such as reduced feed intake, fever, abnormal discharge or unusual behaviour.

- Consult the nearest veterinary aid centre for help if illness is suspected.
- Protect the animals against common diseases.
- In case of outbreak of contagious disease, immediately segregate the sick, in-contact and the healthy animals and take necessary disease control measures. .
- Conduct periodic tests for Brucellosis, Tuberculosis, Johne's disease, Mastitis etc.
- Deworm the animals regularly.
- Examine the faeces of adult animals to detect eggs of internal parasites and treat the animals with suitable drugs.
- Wash the animals from time to time to promote sanitation.

VI. Breeding Care

- Observe the animal closely and keep specific record of its coming in heat, duration of heat, insemination, conception and calving.
- Breed the animals in time.
- The onset of oestrus will be within 60 to 80 days after calving.
- Timely breeding will help achieving conception within 2 to 3 months of calving.
- Breed the animals when it is in peak heat period (*i.e.* 12 to 24 hours of heat).
- Use high quality semen preferably frozen semen of proven sires/bulls.

VII. Care during Pregnancy

- Give special attention to pregnant cows two months before calving by providing adequate space, feed, water etc.

VIII. Marketing of Milk

- Marketing milk immediately after it is drawn keeping the time between production and marketing of the milk to the minimum.
- Use clean utensils and handle milk in hygienic way.
- Wash milk pails/cans/utensils thoroughly with detergent and finally rinse with chloride solution.
- Avoid too much agitation of milk during transit.
- Transport the milk during cool hours of the day.

IX. Care of Calves

- Take care of new born calf.
- Treat/disinfect the navel cord with tincutre of iodine as soon as it is cut with a sharp knife.
- Feed colostrum to calf.

- Assist the calf to suckle if it is too weak to suckle on its own within 30 minutes of calving.
- In case it is desired to wean the calf immediately after birth, then feed the colostrum in bucket.
- Keep the calf separately from birth till two months of age in a dry clean and well ventilated place.
- Protect the calves against extreme weather conditions, particularly during the first two months.
- Group the calves according to their size.
- Vaccinate calves.
- Dehorn the calves around 4 to 5 days of age for easy management when they grow.
- Dispose of extra calves not to be reared/maintained for any specific purpose as early as possible, particularly the male calves.
- The female calves should be properly reared.

7

Business Growth in Dairy Farming

INTRODUCTION

Half of the dairy farm businesses in Australia have less than 200 cows and a gross income under $200,000. Increasing real costs of production and decreasing real price received for milk, and small annual increments in productivity, mean the choice for average sized dairy businesses is to accept low returns or grow.

The aim of this research was to understand the process of growth of dairy farm businesses. Growth is defined as changing the farm system with the aim of increasing net worth in the future.

Growth can also mean 'increased net worth'. To understand the process of growth of a dairy farm business it was necessary to find out about the experiences of some dairy farmers as they expanded their businesses. This was done in the context of theories about firm growth.

Three broad research questions were asked:

1. Why expand a dairy farm business?
2. How does the situation on a dairy farm business affect ensuing business expansion?
3. How does the process and achievement of expansion affect the subsequent operation of a dairy farm business?

METHOD

Five dairy farm businesses were studied. Data was collected using open-ended interviews, direct and indirect observation, and from records. Analysis occurred in two stages. The first stage was analysis within each case.

Eisenhardt said that the overall aim of this step is to define each case as a 'stand alone' entity. It also allows the researcher to see unique patterns in each case. This is helpful when it comes to cross-case analysis. The second stage of analysis was to compare the cases to identify commonalities. From the two-stage analysis of the five case studies a model was derived. In the first part of this model the growth process of a dairy farm business is presented. In this, the distinction between the two types of growth and their affects are highlighted. In the second part of this model is detail about what was involved at specific times and the means used to expand. Here the answers to the questions about how a business/farmer affects ensuing growth, why a farmer expands a business, and how expansion affects growth, are shown.

HOW DOES THE SITUATION ON A DAIRY FARM BUSINESS AFFECT ENSUING BUSINESS EXPANSION?

The state of resources, the size of operation, and the economic and financial situation, all affected growth. The farmers' experience in farming, confidence in their ability, family situation, knowledge, peer support, goals, attitude to current consumption and attitude to risk, affected the opportunities they saw and pursued. Each of the farmers started with a farm that could be improved and stocked more heavily, but their financial situation limited what they could do and how quickly they could act.

PROFIT

Profit and net cash flow of the three farms that expanded, indicated the businesses would be too small in the future to meet needs. The estimated return on assets for these three businesses prior to expansion was less than what could be gained in alternative similarly risky uses of the capital, such as term deposits, government bonds, other property investments and share market investment.

It was also less than larger dairy businesses were earning. The farms had relatively high overhead costs, mainly the cost of labour. Expanding the business would spread this fixed cost over more resources and output.

FINANCIAL RISK

The financial risk these farmers were prepared to accept affected the rate and extent of growth. Farmers A, B and E were prepared to accept increased financial risk. Each reduced their equity in the business to around 20 per cent at the time of major growth. One of the reasons they did this was that they perceived the size of their business put them at more risk of not surviving in the future than did the financial risk from increasing leverage to expand.

Farmer C was the only farmer in this study who was especially wary about expanding 'too fast' and increasing the financial risk. Farmer C had foregone the opportunity to purchase more land. The rate of growth of Farmer C's business was slower than the rate of growth for the other farmers in this study. For Farmers A, B, E the risk of failure because of inability to service debt increased once they had expanded their businesses. Farmer A said at the second interview: You do not get a chance to come back if you over-step the mark – it is not like if you put on too many cows, it is easy to remove the extra cows, rather if you over-borrow it is very difficult to come back. Farmer A would not allow his equity to fall to 20 per cent again. He believed that attitude to equity depended on the stage of farming career and family life.

He now has more capital invested in the business and would only let his equity fall to 50 per cent. Having expanded their businesses, Farmers B and E were not prepared to borrow again to the same extent. This was because they now had more of their own capital to lose. As well, the risk of their business not surviving is lower than before growth. Both the need and desire for growth had diminished.

CONSUMPTION

The decision of a dairy farmer to use more of the annual cash surplus for consumption results in less cash surplus available for debt servicing and investment. This slows the rate equity, or the business, grows.

During expansion, Farmer A, B, D and E lived in old farm houses and invested on farm for greater future consumption, rather than use surplus cash flow for present consumption. The farmers who had gone without a new house and extras in the years prior to growth, wanted to consume more after growth. They felt they had invested for future consumption and that it was now the time for some modest increase in consumption.

INFRASTRUCTURE

The capacity of infrastructure on farm was found to be both a constraint to growth and a motivation for growth. One of the most challenging issues for the farmers prior to, or during, an expansion phase was the question of when to replace capital infrastructure. The farmers considered their financial position and the economics of replacing the dairy. At the time of major growth Farmer B decided they would not replace their existing dairy.

Their existing shed with current employees would be suitable. Two things happened after major growth. First, milking the increased herd took 11 hours a day. These farmers and their employees could barely do the other essential farm work. Farmer B had no time to contemplate other business opportunities. The situation was exacerbated because he had to manually change irrigation throughout the night.

Second, higher than normal milk prices occurred. The unexpected cash surpluses earned from high milk prices meant they were able to pay for half the cost of a replacement dairy. Farmer E replaced the dairy when he was milking 250 cows in a shed and yard designed for 150 cows. Like Farmer B, Farmer E waited until the labour required to operate the older equipment increased and the price of labour increased. Milking more cows than the existing infrastructure efficiently allowed reduced the time Farmer E could put into other business activity. Farmer D replaced their dairy when milking was taking five hours a day, which was more time than initially budgeted. Farmer C also found that the size of the current infrastructure was stopping them from increasing the size of the milking herd at the rate that they wanted. They replaced the infrastructure before they experienced any significant increase in the time of milking the cows, unlike the other three farmers.

This change was instigated by receiving an interest rate subsidy through the Productivity Enhancement Programme of the Rural Finance Corporation. For four out of the five case study farmers, the capacity of the existing capital equipment prevented further growth for a time. In contrast, Farmer A replaced the dairy within the first few years of farming with a dairy that had a much larger capacity than required by the herd at the time.

Consequently the capacity of capital infrastructure became an inducement for further growth for Farmer A. Farmer A was in a different situation to the other four farmers because at the time of replacing the dairy he was a sharefarmer. The owner of the farm business paid for the replacement dairy.The impact that the capacity of capital infrastructure has on business expansion is related to other factors than the dairy. Farmer B installed a labour intensive feed system in the dairy when they first began farming. This feed system involved manually taking the feed to troughs. They fed the cows like this for five years.

With this system it was difficult to feed more cows. Their milk company offered an interest rate subsidy for a loan that was used to upgrade the feed system in the dairy. Comparing the costs of the two systems, the replacement technology was better. Upgrading capital infrastructure changed milking time and the quality of milk. In this study, it was repeatedly seen that when the farmers bought extra land, the whole farm needed improvement. Farmer E for example, found that on all the properties he bought, the fences and other infrastructure were dilapidated and the pastures and the farm layout needed improvement. On these properties the previous farmers had decided not to replace capital infrastructure preceding the time of 'shut down'. Eventually over many years it gets to the stage where all infrastructure needs to be 'replaced'.Farmer E not only had the capital cost of buying the land, but also the capital cost of improving that land. The farmers considered that it would have been more economical if they had started with a 'green-fields' site, rather than becoming a large dairy farm business by buying adjoining but run-down dairy farms.

THE FARMER'S NATURE

Attitudes to risk and business confidence affected the opportunities they saw for business expansion and how they responded. These entrepreneurial characteristics developed over time. For example, at the time of doubling their farm size Farmers A and B planned to expand the herd to 200 or 300 cows – not to 600 or 1000 cows, which was the size these farms eventually became.

It is interesting that each of the farmers in this study spent the first few years of their farming career steadily increasing cow numbers and learning about pasture management and feed management. There appears to be a time period, in the early years of a farmer's career, where farmers develop skills in the technical management of the farm. Early on, the case study farmers spent time dealing with problems and learning from experience.

In this study the farmers themselves linked important events to times where they learnt something important and their confidence improved. In the first few years of farming, Farmer A managed the farm during a run of adverse seasons. This experience, and the success of the management strategy used, built selfbelief in his capabilities and confidence.

Farmers A, B, D and E went beyond their own experiences for ideas for expansion. Farmer A visited large farms in New Zealand that alerted him to new possibilities in dairying. Farmer E used findings from experiments at the local research centre to learn about land-forming and feed management.

Farmer E also learnt from his farm consultant and a progressive group of dairy farmers who were all intent on expanding their businesses. Farmer D read about large farms in the United States of America. The case study farmers gathered information from different settings. They applied new ideas to their farm before they became widespread and the industry 'norm'. Farmer A trialled grain feeding before agricultural research centres and other farmers were using grain.

Farmer E trialled landforming when it was just starting to be investigated at the local research centre. Farmer B adopted grain feeding before it became widely used. This study has started to provide evidence that entrepreneurial abilities of people running businesses that grow, are also continually growing.

THE FARMER AND FAMILY

The farmers who ended up operating large businesses began to expand their business before family issues took precedence. Once the family grew their desire and/or time available to give to expansion was reduced. The goals of his/her family affected growth. The farmers initially wanted to expand to reduce the risk of failure by improving profit; to free up 'management time'; to achieve a better lifestyle for their children; and, for some, to attain a good reputation. Each farmer wanted to be better than others.

Over time the farmer's desire to 'maximise' profit and wealth and security decreased and spending time with the family and fulfilling other goals became more important. These 'non-economic' goals constrained the rate of growth that ensued. Farmer A was looking at decreasing the size of his farm business in the future because his focus was no longer on profit and wealth maximisation. It had changed to improving lifestyle.

EMPLOYEES

The problems with labour that Farmers A, B, D and E experienced, which ultimately resulted in growth being slowed, were:

- The owner-managers having to change and let someone else take over jobs that they previously did;
- Lack of skills in farm employees. This was a more significant problem for farmers who were expanding geographically;
- The time taken to train labour;
- The time taken for labour to be integrated fully into the business and be fully effective;
- Employees were not as motivated or would not work as hard as the farmer. Some of the farmers considered that to achieve better quality and greater motivation in employees they would need to be paid more or offered better conditions than the farmers could afford;
- Employees could leave at any time. These farmers found they could manage if one employee left. It would make the farm difficult to manage at that time but it would be possible. However a number of employees leaving at the same time caused greater problems;
- Employees had difficulty in seeing the 'overall picture'. Farmer B had an excellent farm manager who could do all the jobs on the farm at a very high standard, but he could not be left to manage the farm alone for a period of greater than three weeks or in the crucial times. This farm manager had difficulty in managing the different elements of the system as an integrated farm system;
- Employees had difficulty in setting priorities; and
- Employees often reacted to a problem rather than 'stepping back' and thinking about a solution that will not cause further problems in the future.

Over time, the labour constraint on growth for Farmers A, B and E became less of a problem. Finding skilled employees became easier because these farmers had learnt to train employees. Employees came to them because they had a good reputation within the industry.

They had become employers of choice. Farmers A and B had managers in place and had possible replacement managers in mind in case one of the current managers left. Rewarding and motivating labour became the main labour problem as the businesses expanded. This constraint was managed by offering incentives related to the production of the farm, and by leasing land and/or cows from employees. The employees felt greater 'ownership' of the farm business.

TECHNICAL LIMITS

There was perceived to be a limit to the number of cows that could be milked in a predominantly grazing-based dairy farm, beyond which major diseconomies of size might arise.

The factors which limited the maximum size of the farm included:

- The distance cows could walk per day before time and energy requirements become excessive;
- The quantity of pasture cows can consume at grazing is limited by grazing habit and time and rumen capacity to around a level that will produce 5000 litres per year
- The number of cows that the current infrastructure can handle;
- The number of cows that can be managed logistically. That is, a shed can be used 24 hours a day and the herd could be split. Nevertheless, it is not usually possible to utilise the infrastructure to this extent without the added logistical and time problems increasing costs. Replicating the current farm system on a spatially separate site is a sound option, but this is not straightforward. The two case study farmers who were managing more than one farm and managing a considerable number of people explained that this caused a lot of added management complexities.

The maximum number of farms that can be managed was constrained by the following factors:

- How key individuals within the management team felt about changing their role into a management position with less 'hands-on' work, and giving employees greater responsibility.
- The size of the management team, which constrained how much growth can be contemplated
- Whether the Workforce could manage a larger, more finely-tuned system that required greater understanding of how the parts of the system link

FACTORS OUTSIDE THE FARM BUSINESS THAT AFFECT BUSINESS EXPANSION

It was found that the external environment, which includes financial institutions and the influence of peers affected the expansion that was undertaken.

FINANCIAL INSTITUTIONS

Financial institutions constrained growth. Farmers A, B and E who expanded their businesses rapidly, borrowed to the limit of external financial constraints.Farmer E, who had business skills and performance history could buy only one adjoining dairy farm when he had the opportunity to buy two, because of the external financial constraint.Farmer C, who expanded their business slowly over time, never borrowed to the point where the external financial constraint stopped business expansion. To overcome the external financial limits Farmer A leased land. Another option for farmers to remove the external financial constraint is to recruit an equity partner. However, these farmers were reluctant to recruit an equity partner. Farmer D believed that an outside equity partner would erode his control of the business.

Farmer E would have an equity partner if they had a small share. Farmer A would consider an equity partner if he was going to lease a third farm, but was reluctant because it would increase the number of people involved in decisions. Farmer B would have an equity partner if the circumstances suited and the partner was 'right'. Non-price rationing factors affected the farmers' ability to obtain finance. Several of the farmers mentioned that they had to go to a number of banks before their plans could be financed. Farmer E presented the banks with three plans – one he really wanted to do, one that he was prepared to do if needed, and a readily acceptable plan.

Farmer C used a management consultant to create a business plan when they were buying the home farm and building a house. Farmer B drew on their education and past work experience to help them to prepare cases for loans. To borrow the farmers needed plans, and needed to demonstrate their knowledge and ability to carry out the plan.

PEERS

The influence of peers had both positive and negative affects on business expansion. For the farmers in this study who were managing large farm businesses, meeting managers of larger farms gave them greater knowledge and increased belief in their own ability. Farmer A said 'I discovered that these men and woman managing these large farm enterprises were no different to myself. Such an experience made me realise I can do that too'.

Other positive experiences from peers came from consultants and conferences that inspired them. Two of the case study farmers mentioned that peers could have a negative effect on their decision to expand. Farmer E said My local discussion group has a narrow view and focus on issues that in the 'big picture' are irrelevant.It is only through being a part of a discussion group that includes farmers who all are undertaking major business expansion I am able to learn and to gain confidence that the plan I have for my business is obtainable. Farmer C found the experience of their peers valuable when they were contemplating expanding and employing labour saying: 'other neighbouring farmers have been looking for an employee for 12 months and we are concerned that we could be in the same position'.

HOW DOES EXPANSION AFFECT A DAIRY FARM BUSINESS?

Growth increased the profitability of the businesses studied. Growth affected the skills required of management and labour. Major growth involving significant changes to infrastructure, labour and land in a relatively short time affected the profitability, financial viability and management of the farm business. In contrast, minor growth did not significantly affect management and labour.

MAJOR BUSINESS EXPANSION RESULTED IN ADJUSTMENT COSTS

Immediately following expansion, the farmers of case studies A, B, D and E experienced unplanned adjustment costs. For all these farmers the adjustment costs were in management and livestock. The farmers of case studies A and D, who expanded to a new region, had additional adjustment costs relating to the difficulty of finding and training skilled labour.

Case studies B and E expanded the area of the home farm. They had additional adjustment costs relating to infrastructure and land. In each case the adjustment costs were exacerbated by external factors such as unexpectedly low milk price and unfavourable climatic conditions. During this transition period, where the businesses had to deal with unanticipated adjustment costs, the case study businesses faced increased risk of business failure. This was because the amount of milk produced per cow declinedii at the same time their costs increasediii.

Further, external factors that decreased cash flowiv also reduced the amount of cash surplus available for reinvestment back into the farm. This slowed the gains from growth and exposed these businesses to greater periods of risk. Added to this business risk, each of the farmers had decreased their equity in the business significantly - by up to 30 per cent in a couple of the cases - which consequently increased their financial risk.

THE ADJUSTMENT FOR THE FARMER

The farmer in this study who expanded his business through minor growth had to develop new skills; this farmer had time to adjust and found learning new skills did not slow growth. Farmer C took growth slowly and found the changes required were 'evolutionary'.Pursuing minor growth, Farmer C increased the herd size by a small percentage each year, or took on more adjoining land, and slowly improved that land whilst waiting for their own replacements to utilise that land. The skills required for this growth were needed in the early years of Farmer C's farming career. Farmer C applied those same principles to more cows or more land over time. For minor growth, few new skills are required quickly. Rather, minor growth required fine tuning of technical farm management skills.Farmers A, B, D and E not only had to fine tune technical farm management skills but also had personal adjustment costs immediately after embarking on the major growth phase. As the expansion process was implemented, the work efficiency of each of these farmers decreased as they had to manage unexpected circumstances. Farmers A, B and E all felt, to differing degrees, stress, depression, and some disbelief that the expansion was not delivering all the benefits they had expected.They had expected that by expanding, they would have more time off, would not need to do night work, and would be able to employ more people, which would give them more time with their family. It did not happen this way – not for quite some time. Once time had passed and they had developed the necessary skills, and learnt to accept that others could do their work, and had updated the infrastructure that was needed, then efficiency increased.

The extent growth affects the new skills a farmer requires depends on the rate of growth. Farmer A doubled business size at once, rather than over a few years. The difficulty with rapid expansion was the new knowledge and skills the manager and his employees needed to acquire rapidly. These challenges were exacerbated because the manager on the second farm soon left.

Unlike Farmer A, Farmer E experienced little extra pressure and few difficulties. This was partly because he anticipated the problems, but also because major growth occurred over a number of years.Farmer E bought a property and spent the next two years upgrading it. He employed someone at the start of the major expansion phase who stayed with the business for the entire major expansion phase. Farmer B also purchased a dairy farm that needed improvement. He spent a year improving farm layout and building a new dairy. The slower growth rate for Farmers E and B reduced the rate and extent of personal adjustment required.

THE ADJUSTMENT FOR THE 'MANAGEMENT TEAM'

Growth affected the management team adversely when new personnel were recruited and had to perform well almost immediately. This happened when Farmer A had to find a replacement manager for his second farm when his manager left shortly following expansion.

It also happened when Farmer D took on a third farm and they were relying on finding a person who could quickly develop the skills needed to manage that third farm. Further, when a business gave an employee time to 'develop' within the business before taking on a role as manager, the adjustment costs of expanding the management team were reduced. Farmer E expanded the management team at the same time as he began the four year expansion path. The rate of growth affects the degree of impact the growth has on the management team.

THE AFFECT OF EXPANSION ON THE ECONOMIC AND FINANCIAL STATE OF THE BUSINESS IN THE 'STEADY STATE'

Expansion increased capital efficiency and revenue generation and cost control, and therefore increased the return on capital.Return on assets was increased by increasing operating profit and increasing the asset turnover ratio. Operating profit was increased by spreading fixed costs over more output; by lower prices for inputs through volume purchases; and by higher prices for milk for increased volume. Four of the case study farmers increased their asset turnover by making greater use of resources. Since expanding, all the farmers had increased their net worth.

THE CHANGE IN ROLE OF THE FARMERS

The specific areas of the businesses that were highlighted as changing as a result of major growth were: labour management, financial management, office management, owner/manager's role, technical management and young stock management. Skills in herd management and logistical management also needed to improve. In the following, the requirements of management in each of these areas are described.

LABOUR MANAGEMENT

As these farmers expanded their businesses, their role became more specialised and less directly involved with the daily operations of the farm. There was less physical work and more management work. After a major growth step, Farmers A, B and E were not able to do all the jobs on the farm themselves and had to pass responsibility to others.

All the case study farmers who were managing large farm businesses experienced difficulties caused by the quality of the first farm labourers they employed. When managing larger dairy farming businesses they needed to recruit better calibre employees than their present employees. Each of the case study farmers in this study had strategies for recruiting quality labour.One employed a sharefarmer, another a family friend, another promoted a farm labourer to manager and employed an existing employee to be manager on the home farm. The farmers also employed contractors for specific tasks.

The farmers found they needed good negotiating skills and communication skills in order to manage labour successfully. Building strong relationships with the managers was important; involving them in decision making and encouraging them to attend discussion groups and to undertake further training was also important.

FINANCIAL MANAGEMENT

The dairy businesses with high equity and around the average size had rudimentary requirements for financial management, compared to the expertise needed for the larger operations. Financial management became more important as the farmers expanded the business. The farmers had borrowed and reduced their equity percentage to finance growth. Farmer B, for example, reduced the equity percentage from 50 per cent to 20 per cent in order to expand the business.It was at this time that financial risk was great, and the farmers needed to monitor their cash flows closely, and to enhance skills in managing their finances. High gearing ratios meant the farmers needed high-level negotiating skills in the financially difficult years. Farmer A explained that it was important to communicate regularly with the people who had a major share in the business to let them know what was happening during the difficult years.

OFFICE MANAGEMENT

For a farmer managing 100 to 200 cows, the extent of office management was writing cheques, with some cash flow control. As a farm expanded, office management became a greater proportion of the farmers work and the farm became a more complex business to manage.

Office management on a large farm involved:

- Researching new ideas;
- Organising the jobs that needed to be done and setting priorities;
- Organising time off for employees;
- Planning what jobs were needed on the farm;
- Analysing and dissecting the events on the farm;
- Implementing decisions;
- Paper work associated with employing labour;
- Financial management;
- Keeping track of the set goals (one of the farmers said that it was very easy to become disillusioned when you have such a huge debt. He said that you almost feel as though you are working to pay the bank back. But he said the key was to keep your mind on the end goal); and
- Juggling the different demands of the farm.

The time available to devote to office management was significantly less than the farmers had anticipated. Thus, time management was another skill these farmers had to develop.

The farmers dealt with issues of time management by:

- Delegating jobs to their employees;
- Using contractors (however, they found that they needed to build up a relationship with contractors so that they would show up when they said they would and so that they did not need to be there to supervise the contractors work.); and
- Being flexible on the number of days spent in the office and the number of days on farm. One of the farmers, who expanded to another spatially separate farm, spent six months planning for growth. This involved thinking through what changes needed to be done on the property, the labour required and his own change in role. Farmer E

explained that a lot of planning went into replacing the dairy. Farmer D spent a lot of time planning new business activities. Farmer B spent every Wednesday planning.

OWNERS ROLE

The owner's role on a 500 plus cow farm involved spending some days in the office and some days helping on the farm. Farmer E described himself as filling in when someone is away or helping at the busy times on the farm.

This gave more time doing other activities, such as increased involvement with research bodies. Farmer B also found that after the initial phase expansion gave time to explore other ventures.

TECHNICAL MANAGEMENT

Technical management – the management of the pastures, the herd, feeding, and so on was crucial no matter what size the business.However, as the business expanded, technical management became more complex and difficult. This was partly because the farm was on a much larger scale, which meant that problems were amplified. It also meant new problems and complexities evolved.

YOUNG STOCK MANAGEMENT

The farmers found that to expand the business they had to remove young stock from the home property. They found that the land adjoining their property was best used for their milking herd. They either had to contract rear their young stock or they had to lease/buy a run-off block.

HERD MANAGEMENT

With around 100 to 200 cows, the farmers knew cows individually. As the number of cows in the herd increased the management of the herd changed. Before expansion, the farmers did not fully appreciate the added work specifically associated with increasing stock numbers. The farmers had considered the implications of milking more cows and the infrastructure requirements. However, they generally had not considered sufficiently the detailed logistics associated with moving more cows, joining more cows, feeding more cows, calving more cows, looking after more dry stock and rearing more young stock.The additional problems associated with managing more cows related to wear on the tracks, more cows to fit into the yards, more cows to go through the milking shed, and more cows to obtain access to water. These factors had major implications for feed management, the use of out-paddocks, contract rearing and so forth.

LOGISTICS MANAGEMENT

The degree of improvement required in logistical management skills depended on the rate of major growth. The dairy farmers in this study who rapidly doubled their herd size did not have sufficient time to develop adequately the required logistical management skills for managing a herd of twice the original size, and experienced additional problems for this reason.

Expanding the area of their home farm rapidly without replacing the old dairy caused further logistical problems. Some of the farmers at the time of expanding did not have the capital to replace their existing shed. They also doubted whether it would be economically sensible to replace the shed at that time.

They believed that by dividing the herd into two, and then having two milkings in the morning and two milkings in the evenings, the current shed would be a viable solution. However, the logistics associated with organising two herds and the time involved meant that this 'solution' was inefficient and the farmers ended up spending much more of their day in the dairy shed than planned.

The dairy farmers that expanded their businesses geographically experienced added complexity with managing the logistics of the human resources. These farmers had more people to manage, which were on two spatially separate farms.

8

Career in Dairy Farming

DAIRY FARMING

If the advertisement is to be believed and it really takes one glass of milk to produce one slice of cheese, one can only imagine how many glasses of milk are needed, not only to provide the thousands of slices of cheese that are being readily gobbled up by India's yuppy children, but also for regular consumption and all the other milk derived goodies. The demand for milk is tremendous, and is growing not only in cities but also in small towns and rural areas.

Dairying plays a dynamic role in India's agro-based economy. Our country's Dairy Industry is one of the fastest expanding in the world. Ten years ago only 5% of the milk produced came into the dairies. Today it is 10%, and this figure will keep increasing. Today India ranks second only to the US in terms of milk production, and it is expected that at the present rate of growth, will overtake the US by the Year 2000.

A recent report prepared by the Ministry of Agriculture, the dairy industry has the potential to offer about 4.2 crore jobs a year. Today there are barely 50,000 – 60,000 trained professionals in this field. Given the proper thrust, dairying could in the years ahead, become one of the largest business sectors of the economy, even overtaking the industrial sector.

There are over 200 dairies spread across the country providing us around 60 million tonnes of milk. But this only constitutes the organised sector, which includes the dairy co-operatives, Government run milk producing units and the private sector dairies, and not the vast unorganised sector, where the milk is delivered directly at the door, and does not go through a dairying process.

Over the past 35 years, Dairy Farming has grown from this largely unorganised – if complex – activity, into a vast organised industry, that encompasses not only increased production of milk and milk products, but also the breeding of higher yielding cattle, and the scientific rearing of cattle and production of cattle feeds. This has proportionately increased the demand for trained manpower in this sector.

WORK

The two main areas of activity in the Dairy Industry include:

- Liquid Milk Production which covers its collection, processing and distribution. This is carried out almost entirely in the public sector through cooperatives.
- And the Manufacture of Milk Products: like butter, cheese, ghee, condensed and evaporated milk, dried and skimmed milk, malted foods, chocolates and ice cream – where value addition ensures profitability which is largely undertaken by private companies.

Most jobs in this industry fall under two broad heads:

- Production, where work is mainly with milch cattle and includes the most important task that of obtaining milk, as also breeding of high yielding cattle and taking care of the animals. Veterinary Scientists are usually best suited for such jobs. They are appointed as Procurement Officers and must visit villages talk to farmers and influence them to sell their milk. Dairy scientists in the area of milk production also conduct experiments to determine the effects of different kinds of feeds and environmental conditions on the quantity, quality and nutritive value of milk produced. They also research breeding, feeding and management of dairy cattle.
- Processing, as the term implies, covers the handling of milk for distribution; or its conversion into dairy products. This work normally begins after the milk has reached the plant. Here milk is converted into a variety of dairy products, such as cheese, chocolates, ice cream, butter ghee and so on. Dairy Technologists deal with this aspect of the industry, and also work to develop improved methods in production preservation and utilisation of milk, and milk products.
- Other professionals required in this specialised area include dairy engineers, who are responsible for the setting up and maintenance of dairy plants and related activities; and marketing personnel who handle the marketing and sales of milk and milk products.

ENTRY

Specialisation in the specific or related area, like Veterinary Science, Dairy Technology or Engineering is a pre-requisite for entry into this field.

ELIGIBILITY

In general, most courses require candidates to have passed the 10 + 2 examination, with Chemistry, Physics, Mathematics and English and with a minimum aggregate of 55%.After graduating in dairying, candidates can opt for further studies and specialisation through a Master" programme in Dairying, or related areas such as, genetics and breeding, livestock production, Animal bio-technology or in fields of food technology and preservation.

SELECTION

Selection is on the basis of an All India Entrance Examination.

DURATION

The B.Tech/B.Sc in Dairy Technology is a four years course after 10+2. The B.VSc, is also a four years programme after 10+2, and the Dairy Diploma course of two years duration for lower level technical positions.

INSTITUTIONS

Some of the prominent institutions offering courses in dairy technology include:

- Dairy Science College, National Dairy Research Institute, Karnal, Harayana. National Dairy Research Institute, Bangalore.
- Sheth MC College of Dairy Sciences, Anand, Gujarat Dairy Science Institute, Aarey Milk Colony, Bombay
- College of Dairy Sciences, Udaipur, Rajasthan Veterinary College, Hebbal, Karnataka.

PERSONAL ATTRIBUTES REQUIRED

While dairy farming can be a challenging career offering enormous prospects for trained professionals, those getting into this field must be prepared to live in remote areas, or far away from the comforts and conveniences of the city, which could place pressures on those unprepared for this sort of lifestyle. It also means long hours of very hard work.

Moreover, as milk is a perishable commodity – the pressures involved in maintaining product quality standards, can be tremendous.

JOB PROSPECTS

Openings for Dairy Technologists are available in both the public and the private sector. Co-operatives and federations provide ample scope for the Dairy Technologists in the public sector. In the private sector, specialists can join rural banks or private enterprises that manufacture and market milk products. A large number of Dairy Technologists also start up their own businesses in milk production or processing or work as consultants.

Research and development is another growing area, crucial for the dairy industry.

CAREERS IN DAIRY PRODUCTION

It is a pleasure to address the Student Section of the American Dairy Science Association and discuss briefly with you Careers in Dairy Production. There is no subject closer to my heart. As a dairy farmer and cattle breeder, as a professor of Dairy Production for nearly 25 years, and with some eight years of experience in my present position, I have formed some very definite opinions regarding the future. The ideas I present are gathered from experience and observation over the entire dairy production field.

FUTURE NEVER MORE CHALLENGING

In my opinion, the future for a worthwhile career in dairy production was never more promising. At the same time, I am sure the man who is successful in this field in the future must be better prepared from the standpoint of knowledge, Judgement, business sense, and courage than the dairy production major who graduated twenty years ago. This is largely because of the tremendous changes in our agricultural picture in the past twenty years. The dairy production picture has changed a great deal.

As you know, in 1945 we had nearly five million farmers. Now we have only about three and one-third million. We had well over one million dairy farms selling milk or cream- now we have only about 985,000. Our average farm is larger by 100 acres. We have 25% less milk cows and some 25% fewer dairy farmers. In 1916, our average DHIA-tested herd was 16 cows. Now it is 44 head and still growing.

The producer-distributor has almost disappeared. So have many of our small creameries. Our local cooperative creameries so vital to the dairy farmer 20 to 30 years ago have merged, joined with, or sold out, to a national dairy operation, or have folded, up. In this milk business we have developed volume operations that result in the cheapest and best bottle of milk in the world, delivered to our consuming public.

We have moved into an area where service and selling, as well as producing and processing, play a vital part. At the same time, we have increased emphasis upon a quality product. In the milk-producing business we would not want the quality picture otherwise.

We also have come to recognize the dairy business, both production and products fields, as part of a highly developed and highly competitive field.

It is a business that requires more capital investment per worker than any phase of agriculture and, as you know, the investment per average farm worker is $27,000 and in industry it is only about $15,000 per man. I can cite you successful dairy production operations where the investment per worker is in the area of $30,000 to $50,000 per man-land, cattle, buildings, equipment, etc.

Dairymen who recognize this challenge and meet it are making a living and a fair return on the investment. Any man who desires a career in dairy production can succeed if he has the correct perspective of the job ahead. I should be the last person in the world to criticize our teaching in agricultural colleges, but I do not hesitate to say that in my opinion There is often too much attention and time spent on minute details and the student never gets a glimpse of the broad picture he faces when he graduates.

Frankly, I feel that any student who wishes to succeed in this field should bury his diploma good and deep, not haggle too much about starting salary and time off, and get a few years of good hard-headed experience-if he hasn't had it before-and then start to build his career. He then will find his college training invaluable and, along with good work habits, his career will begin to unfold.

WHAT ARE SOME OF THE JOBS ?

Time and space will not permit my listing any but a few key jobs in a few fields. Keep in mind that as a rule jobs go to those best prepared to fill them! By the same token in merchandising—even our artificial insemination of cattle programme of which I am very proud—"markets go to those best prepared to serve them."

Dairy Farming and Cattle Breeding

For the dairy production major, the love of cattle and of the land is usually very strong. Even though he finds himself working for a salary, he dreams of the day he can own his farm and his herd. This goal is attained sometimes, but not too often, for the simple reason it takes so much money to start dairy farming the average man, after paying income tax, supporting a family, owning a home and a car, hasn't much left.

The best way today to begin, for a young man who wants to be a dairy farmer, is to go right back home and go into partnership with his father and put his ambition and knowledge to work.

Another plan is to go into partnership with an old, established dairy farmer and gradually buy into the setup. Most of our dairy farmers are past 50 years old. Some very good herd owners are looking for the young man who loves cattle and the soil, to team up with.

Such jobs usually start with the young man acting as assistant manager. If he has the stuff he gradually becomes part-owner. There are more opportunities along this line than many people appreciate. Just consider the established herds in your own community. Who will carry on when the present owner no longer can"hit the ball"? (e) along with most established herds is the purebred cattle challenge and the pleasure of producing some seed stock, as well as milk. Today, we have too few constructive breeders.

We need in our a.i. programme more proved sires, more good cow families and, industry wise, more good cattle. Due to our present income tax laws, and our farm labour picture, there is a declining number, and will continue to be, of wealthy herd-owner hobbyists. These men, by virtue of sufficient capital, assembled many great animals.

They bred some of our best cattle. They also purchased some of the top cattle from smaller breeders. In the future, farm-sized herds and the semi-commercial herd will, by virtue of a sound breeding and testing programme, produce much of our seed stock for the future. The farm-sized herd, where some individual cow attention can be given, will continue to be the leader in this area.Going into business for himself will not for a long time have the glitter of a salary job. New cars, new suits, and a few of those things won't come too fast. An understanding wife is a must in this picture. Most fellows who choose this course won't be quitting at 5 p.m. And come in smelling like a rose for a 6 p.m. Dinner.

In the long run, however, they will amass more total capital, have more independence, more freedom of action and thought, be healthier, have a closer-knit family life, be more appreciated in the community, and have less heart trouble than the classmate who chooses the humdrum city life and keeping up with the joneses.

Herd Management Farm Superintendent

It has long been said that "If you can't own your own herd the next best thing is to manage a good herd and farm for someone else." This is true today—but different than 25 years ago. There are fewer farms supported by some other business. Most farms today have to make money or at least pay interest on the investment. In America, there are vastly fewer jobs of estate managers than we once knew.

There will continue to be less! The hired dairy farm manager today will have to: Possess good business Judgement and be able to handle large capital and to direct employees.

He will have to be an expert dairyman, as well as a farmer, and produce milk in large volume at low cost to meet competitive market demands. He will have to possess ingenuity and exert energy to keep his operation right-side-up in the face of the changing agricultural picture. He will also have to be a good salesman, to insure his employees and his stockholders that the future is sound and that his operations are equally sound.

When one works for a Board or a corporation, he must always be prepared to defend his position. It is my observation that to some extent good cow men and men capable of large-scale dairy operations are born with much of the potential.

I have never believed it can be taught except by experience. I do believe sound teaching plus good experience can develop many men along these lines. In the years ahead or see an acute shortage of capable managers who can handle a $500,000 to a two million-dollar dairy production setup. This is what we must face, because as suggested, have many more 200 to 2,500 cow herds, with the resulting capital, labour, feed, disease, and marketing problems. Men who can handle these jobs will always be few, but well paid. They should be well paid, as they are virtually, in the dairy field, what the economists call captains of industry.

Dairy Procurement and Field Work

There will continue to be increasing emphasis upon much procurement. As our dairy processes become fewer, with greater volume, greater overhead expenses, and tougher competition the importance of milk procurement increases.

Plants are constructed to handle milk. They cost a lot of money. Without sufficient volume they quickly become a liability. Procurement work involves direction of field staff, quality control, truck routes, and a score of problems that challenge the best of every worker in this field. It requires from the director to the newest fieldman a thorough understanding and appreciation of the problems of the producer and willingness to cooperate with him.

Dairy Procurement and Field Work

Our breed registry organizations are constantly adding men to serve as editors, advertising representatives, and field representatives in charge of registration and production testing. It is a good field and offers an opportunity to work with people and cattle.

Artificial insemination: This field is new. Starting just 21 years ago, the programme now embraces over seven million, or 37%, of our milk cows in America. There are roughly 12,000 employees in this field at present, but the number increases daily.

Here are a few of the key jobs and their requirements:

- *General manager*: Requires a man capable of directing from 50 to 1,000 employees and heading a business with gross returns of $300,000 to as much as five million dollars a year. General Managers must be interested in cattle improvement, understand business management and finance, be good personnel directors, and be able to carry on a strong public relations programme.
- *Assistant manager*: Assists the manager in his varied duties. Requires same type of experience and training. Offers advancement to full managership.
- *Field supervisor*: In charge of field operations. Includes supervision and working with technicians, selling the programme, public relations and, in some cases, trouble shooting. This is a type of extension work in many ways. Large organizations have a Director of Field Service, who directs the work of many field supervisors. Love of cattle and ability to supervise personnel is a must in this position.
- *Sire analyst*: Every organization has a sire analyst, whose duties are to constantly evaluate sires in use and comb the field for improved sires that will improve the patron's herds. One or more assistants may also be employed. This job requires a basic understanding of animal genetics, working knowledge of breed families, ability to

analyse records, evaluate cattle for conformation, etc., and prepare copy for advertising sires in use.

- *Laboratory supervisor*: This man must be scientifically trained. He has charge of semen collection, processing, storage, and shipping. His position is similar to that of Production Manager in many industries. A key job and one of the most vital. A man in this position must be research-minded, because he is constantly ferreting out weaknesses in his organization's operations that influence settling rate in the field.
- *Technician or inseminator*: While mentioned last, by no means is the technician, whose job it is to inseminate cows, the least important in this field. In most eases he is the key man. He is usually well paid—if he is in a heavily populated cow area and wants to work. Many technicians are college graduates, though the majority are not. Working as a technician for a year or two is vital to the success of men in all positions in A.I. work. This is the place a college graduate should start his training for a more advanced position in this field.
- *Other jobs*: Most organizations publish a monthly or quarterly pamphlet, sent to all members. A part-time, or in some cases a% full-time, editor is employed. He also assists with publicity, sire catalogs, advertising, and public relations. Here is the field for the dairy production man who likes to write.

The National Association of Artificial Breeders is the overall organization combining the A.I. organizations in America. It also has members in Canada and numerous foreign countries. This organization is set up to carry on education, research, promotion, public relations, and aid in regulation of the programme. Positions on the NAAB Staff include Executive Secretary, Assistant Secretary, Advertising Manager, Editor, etc.

Teaching Extension-and Agricultural Experiment Station Research

The immense opportunities for service to the entire dairy production field through teaching, and by extension work in all of its phases, is known to most agricultural college graduates. *It is a worthy field.* Men who expect to teach Dairy Production should be well fortified by actual experience in the production field and have a broad perspective of the field.

Research is vital to our continued progress. Our research men today must have much more fundamental science than was true 30 to 40 years ago. The opportunities are great and the young man who expects to enter the research field would do well to take considerable post-graduate work in basic biological, physiological, and physical sciences.

Industry

The dairy industry claims most of our production graduates. The openings are too numerous to mention. Fundamentally, the ability to sell, advertise, write, and speak well are assets for all except key research men in these fields concerned with merchandising our nearly 130 billion pounds of milk a year. Technically trained men in quality control work, dairy and farm equipment, plant foods, insecticides, building materials are used throughout the industry.

9

Biotechnology as a Career in Dairy Farming

AN OVERVIEW

Biotechnology is a field of applied biology that involves the use of living things in engineering, technology, medicine, and other useful applications. Modern use similar term includes genetic engineering as well as cell- and tissue culture technologies. The concept encompasses a wide range of procedures for modifying living organisms according to human purposes - going back to domestication of animals, cultivation of plants, and "improvements" to these through breeding programmes that employ artificial selection and hybridization. By comparison to biotechnology, bioengineering is generally thought of as a related field with its emphasis more on higher systems approaches for interfacing with and utilizing living things. The United Nations Convention on Biological Diversity defines biotechnology as:

"Any technological application that uses biological systems, living organisms, or derivatives thereof, to make or modify products or processes for specific use."

Biotechnology draws on the pure biological sciences and in many instances is also dependent on knowledge and methods from outside the sphere of biology. Conversely, modern biological sciences are intimately entwined and dependent on the methods developed through biotechnology and what is commonly thought of as the life sciences industry. Biotechnology is not limited to medical/health applications.

Although not normally thought of as biotechnology, agriculture clearly fits the broad definition of "using a biotechnological system to make products" such that the cultivation of plants may be viewed as the earliest biotechnological enterprise. Agriculture has been theorized to have become the dominant way of producing food since the Neo-lithic Revolution.

The processes and methods of agriculture have been refined by other mechanical and biological sciences since its inception. Through early biotechnology, farmers were able to select the best suited and crops having the highest yield to produce enough food to support a growing population. Other uses of biotechnology were required as crops and fields became increasingly large and difficult to maintain. Specific organisms and organism byproducts were used to fertilize, restore nitrogen, and control pests.

Throughout the use of agriculture, farmers have inadvertently altered the genetics of their crops through introducing them to new environments and breeding them with other plants—one of the first forms of biotechnology. Cultures such as those in Mesopotamia, Egypt, and India developed the process of brewing beer. It is still done by the same basic method of using malted grains to convert starch from grains into sugar and then adding specific yeasts to produce beer.

In this process the carbohydrates in the grains were broken down into alcohols such as ethanol. Ancient Indians also used the juices of the plant Ephedra vulgaris and used to call it Soma. Later other cultures produced the process of Lactic acid fermentation which allowed the fermentation and preservation of other forms of food. Fermentation was also used in this time period to produce leavened bread. Although the process of fermentation was not fully understood until Pasteur's work in 1857, it is still the first use of biotechnology to convert a food source into another form. For thousands of years, humans have used selective breeding to improve production of crops and livestock to use them for food.

In selective breeding, organisms with desirable characteristics are mated to produce offsprings with the same characteristics. For example, this technique was used with plants to produce largest and sweetest corn in order to produce the most desirable crops.

In the early twentieth century scientists gained a greater understanding of microbiology and explored ways of manufacturing specific products. In 1917, Chaim Weizmann first used a pure microbiological culture in an industrial process, that of manufacturing corn starch using Clostridium acetobutylicum, to produce acetone, which the United Kingdom desperately needed to manufacture explosives during World War I.

Biotechnology has led to the development of antibiotics. In 1928, Alexander Fleming discovered the mold Penicillium. His work led to the discovery and purification of the antibiotic penicillin. In 1940, penicillin became available for medicinal use to treat bacterial infections in humans.

The field of modern biotechnology is thought to have largely begun on June 16, 1980, when the United States Supreme Court ruled that a genetically modified microorganism could be patented in the case of Diamond v. Chakrabarty. Indian-born Ananda Chakrabarty, working for General Electric, had developed a bacterium capable of breaking down crude oil, which he proposed to use in treating oil spills. Revenue in the industry is expected to grow by 12.9% in 2008. Another factor influencing the biotechnology sector's success is improved intellectual property rights legislation—and enforcement—

worldwide, as well as strengthened demand for medical and pharmaceutical products to cope with an ageing, and ailing, U.S. population. Rising demand for biofuels is expected to be good news for the biotechnology sector, with the Department of Energy estimating ethanol usage could reduce U.S. petroleum-derived fuel consumption by up to 30% by 2030. The biotechnology sector has allowed the U.S. farming industry to rapidly increase its supply of corn and soybeans—the main inputs into biofuels—by developing genetically modified seeds which are resistant to pests and drought. By boosting farm productivity, biotechnology plays a crucial role in ensuring that biofuel production targets are met.

APPLICATIONS

Biotechnology has applications in four major industrial areas, including health care, crop production and agriculture, Non- food uses of crops and other products, and environmental uses.

For example, one application of biotechnology is the directed use of organisms for the manufacture of organic products. Another example is using naturally present bacteria by the mining industry in bioleaching. Biotechnology is also used to recycle, treat waste, clean up sites contaminated by industrial activities, and also to produce biological weapons.

A series of derived terms have been coined to identify several branches of biotechnology, for example:

- Bioinformatics is an interdisciplinary field which addresses biological problems using computational techniques, and makes the rapid organization and analysis of biological data possible. The field may also be referred to as computational biology, and can be defined as, "conceptualizing biology in terms of molecules and then applying informatics techniques to understand and organize the information associated with these molecules, on a large scale." Bioinformatics plays a key role in various areas, such as functional genomics, structural genomics, and proteomics, and forms a key component in the biotechnology and pharmaceutical sector.
- Blue biotechnology is a term that has been used to describe the marine and aquatic applications of biotechnology, but its use is relatively rare.
- Green biotechnology is biotechnology applied to agricultural processes. An example would be the selection and domestication of plants via micropropagation. Another example is the designing of transgenic plants to grow under specific environments in the presence of chemicals. One hope is that green biotechnology might produce more environmentally friendly solutions than traditional industrial agriculture. An example of this is the engineering of a plant to express

a pesticide, thereby ending the need of external application of pesticides. An example of this would be Bt corn. Whether or not green biotechnology products such as this are ultimately more environmentally friendly is a topic of considerable debate.

- Red biotechnology is applied to medical processes. Some examples are the designing of organisms to produce antibiotics, and the engineering of genetic cures through genetic manipulation.
- White biotechnology, also known as industrial biotechnology, is biotechnology applied to industrial processes. An example is the designing of an organism to produce a useful chemical. Another example is the using of enzymes as industrial catalysts to either produce valuable chemicals or destroy hazardous/polluting chemicals. White biotechnology tends to consume less in resources than traditional processes used to produce industrial goods. The investment and economic output of all of these types of applied biotechnologies is termed as bioeconomy.

Medicine

In medicine, modern biotechnology finds promising applications in such areas as:

- Drug production
- Pharmacogenomics
- Gene therapy
- Genetic testing: techniques in molecular biology detect genetic diseases. To test the developing fetus for Down syndrome, Amniocentesis and chorionic villus sampling can be used.

Pharmacogenomics

Pharmacogenomics is the study of how the genetic inheritance of an individual affects his/her body's response to drugs. It is a coined word derived from the words "pharmacology" and "genomics". It is hence the study of the relationship between pharmaceuticals and genetics. The vision of pharmacogenomics is to be able to design and produce drugs that are adapted to each person's genetic makeup.

Pharmacogenomics results in the following benefits:

- Development of tailor-made medicines. Using pharmacogenomics, pharmaceutical companies can create drugs based on the proteins, enzymes and RNA molecules that are associated with specific genes and diseases. These tailor-made drugs promise not only to maximize therapeutic effects but also to decrease damage to nearby healthy cells.
- More accurate methods of determining appropriate drug dosages.

Knowing a patient's genetics will enable doctors to determine how well his/her body can process and metabolize a medicine. This will maximize the value of the medicine and decrease the likelihood of overdose.

- Improvements in the drug discovery and approval process. The discovery of potential therapies will be made easier using genome targets. Genes have been associated with numerous diseases and disorders. With modern biotechnology, these genes can be used as targets for the development of effective new therapies, which could significantly shorten the drug discovery process.
- *Better vaccines*: Safer vaccines can be designed and produced by organisms transformed by means of genetic engineering. These vaccines will elicit the immune response without the attendant risks of infection. They will be inexpensive, stable, easy to store, and capable of being engineered to carry several strains of pathogen at once.

Pharmaceutical Products

Most traditional pharmaceutical drugs are relatively simple molecules that have been found primarily through trial and error to treat the symptoms of a disease or illness. Biopharmaceuticals are large biological molecules known as proteins and these usually target the underlying mechanisms and pathways of a malady; it is a relatively young industry. They can deal with targets in humans that may not be accessible with traditional medicines. A patient typically is dosed with a small molecule via a tablet while a large molecule is typically injected.

Small molecules are manufactured by chemistry but larger molecules are created by living cells such as those found in the human body: for example, bacteria cells, yeast cells, animal or plant cells.

Modern biotechnology is often associated with the use of genetically altered microorganisms such as E. coli or yeast for the production of substances like synthetic insulin or antibiotics. It can also refer to transgenic animals or transgenic plants, such as Bt corn. Genetically altered mammalian cells, such as Chinese Hamster Ovary cells, are also used to manufacture certain pharmaceuticals. Another promising new biotechnology application is the development of plant-made pharmaceuticals.

Biotechnology is also commonly associated with landmark breakthroughs in new medical therapies to treat hepatitis B, hepatitis C, cancers, arthritis, haemophilia, bone fractures, multiple sclerosis, and cardiovascular disorders. The biotechnology industry has also been instrumental in developing molecular diagnostic devices that can be used to define the target patient population for a given biopharma-ceutical. Herceptin, for example, was the first drug approved for use with a matching diagnostic test and is used to treat breast cancer in women whose cancer cells express the protein HER2.Modern biotechnology can be used to manufacture existing medicines relatively easily and cheaply. The first genetically engineered products were medicines designed to treat human diseases.

To cite one example, in 1978 Genentech developed synthetic humanized insulin by joining its gene with a plasmid vector inserted into the bacterium Escherichia coli. Insulin, widely used for the treatment of diabetes, was previously extracted from the pancreas of abattoir animals. The resulting genetically engineered bacterium enabled the production of vast quantities of synthetic human insulin at relatively low cost.

In 2003 study undertaken by the International Diabetes Federation on the access to and availability of insulin in its member countries, synthetic 'human' insulin is considerably more expensive in most countries where both synthetic 'human' and animal insulin are commercially available: *e.g.* within European countries the average price of synthetic 'human' insulin was twice as high as the price of pork insulin. Yet in its position statement, the IDF writes that "there is no overwhelming evidence to prefer one species of insulin over another" and "[modern, highly purified] animal insulins remain a perfectly acceptable alternative.

Modern biotechnology has evolved, making it possible to produce more easily and relatively cheaply human growth hormone, clotting factors for hemophiliacs, fertility drugs, erythropoietin and other drugs. Most drugs today are based on about 500 molecular targets. Genomic knowledge of the genes involved in diseases, disease pathways, and drug-response sites are expected to lead to the discovery of thousands more new targets.

Genetic Testing

Genetic testing involves the direct examination of the DNA molecule itself. A scientist scans a patient's DNA sample for mutated sequences. There are two major types of gene tests. In the first type, a researcher may design short pieces of DNA whose sequences are complementary to the mutated sequences. These probes will seek their complement among the base pairs of an individual's genome.

If the mutated sequence is present in the patient's genome, the probe will bind to it and flag the mutation. In the second type, a researcher may conduct the gene test by comparing the sequence of DNA bases in a patient's gene to disease in healthy individuals or their progeny.

Genetic testing is now used for:

- Carrier screening, or the identification of unaffected individuals who carry one copy of a gene for a disease that requires two copies for the disease to manifest;

- Confirmational diagnosis of symptomatic individuals;
- Determining sex;
- Forensic/identity testing;
- Newborn screening;
- Prenatal diagnostic screening;
- Presymptomatic testing for estimating the risk of developing adult-onset cancers;
- Presymptomatic testing for predicting adult-onset disorders.

Some genetic tests are already available, although most of them are used in developed countries. The tests currently available can detect mutations associated with rare genetic disorders like cystic fibrosis, sickle cell anemia, and Huntington's disease.Recently, tests have been developed to detect mutation for a handful of more complex conditions such as breast, ovarian, and colon cancers. However, gene tests may not detect every mutation associated with a particular condition because many are as yet undiscovered, and the ones they do detect may present different risks to different people and populations.

Controversial Questions

The absence of privacy and anti-discrimination legal protections in most countries can lead to discrimination in employment or insurance or other misuse of personal genetic information.

This raises questions such as whether genetic privacy is different from medical privacy:

- *Reproductive issues*: These include the use of genetic information in reproductive decision-making and the possibility of genetically altering reproductive cells that may be passed on to future generations. For example, germline therapy forever changes the genetic make-up of an individual's descendants. Thus, any error in technology or Judgement may have far-reaching consequences. Ethical issues like designer babies and human cloning have also given rise to controversies between and among scientists and bioethicists, especially in the light of past abuses with eugenics.
- *Clinical issues*: These centre on the capabilities and limitations of doctors and other health-service providers, people identified with genetic conditions, and the general public in dealing with genetic information.
- *Effects on social institutions*: Genetic tests reveal information about individuals and their families. Thus, test results can affect the dynamics within social institutions, particularly the family.
- Conceptual and philosophical implications regarding human responsibility, free will vis-à-vis genetic determinism, and the concepts of health and disease.

Gene Therapy

Gene therapy may be used for treating, or even curing, genetic and acquired diseases like cancer and AIDS by using normal genes to supplement or replace defective genes or to bolster a normal function such as immunity.

It can be used to target somatic or gametes cells. In somatic gene therapy, the genome of the recipient is changed, but this change is not passed along to the next generation. In contrast, in germline gene therapy, the egg and sperm cells of the parents are changed for the purpose of passing on the changes to their offspring.

There are basically two ways of implementing a gene therapy treatment:

- *Ex vivo, which means "outside the body"*: Cells from the patient's blood or bone marrow are removed and grown in the laboratory. They are then exposed to a virus carrying the desired gene. The virus enters the cells, and the desired gene becomes part of the DNA of the cells. The cells are allowed to grow in the laboratory before being returned to the patient by injection into a vein.
- *In vivo, which means "inside the body"*: No cells are removed from the patient's body. Instead, vectors are used to deliver the desired gene to cells in the patient's body.

As of June 2001, more than 500 clinical gene-therapy trials involving about 3,500 patients have been identified worldwide. Around 78% of these are in the United States, with Europe having 18%.

These trials focus on various types of cancer, although other multigenic diseases are being studied as well. Recently, two children born with severe combined immunodeficiency disorder were reported to have been cured after being given genetically engineered cells.

Gene therapy faces many obstacles before it can become a practical approach for treating disease.

At least four of these obstacles are as follows:

1. *Gene delivery tools*: Genes are inserted into the body using gene carriers called vectors. The most common vectors now are viruses, which have evolved a way of encapsulating and delivering their genes to human cells in a pathogenic manner. Scientists manipulate the genome of the virus by removing the disease-causing genes and inserting the therapeutic genes. However, while viruses are effective, they can introduce problems like toxicity, immune and inflammatory responses, and gene control and targeting issues. In addition, in order for gene therapy to provide permanent therapeutic effects, the introduced gene needs to be integrated within the host cell's genome. Some viral vectors effect this in a random fashion, which can introduce other problems such as disruption of an endogenous host gene.

2. *High costs*: Since gene therapy is relatively new and at an experimental stage, it is an expensive treatment to undertake. This explains why current studies are focused on illnesses commonly found in developed countries, where more people can afford to pay for treatment. It may take decades before developing countries can take advantage of this technology.
3. *Limited knowledge of the functions of genes*: Scientists currently know the functions of only a few genes. Hence, gene therapy can address only some genes that cause a particular disease. Worse, it is not known exactly whether genes have more than one function, which creates uncertainty as to whether replacing such genes is indeed desirable.
4. *Multigene disorders and effect of environment*: Most genetic disorders involve more than one gene. Moreover, most diseases involve the interaction of several genes and the environment. For example, many people with cancer not only inherit the disease gene for the disorder, but may have also failed to inherit specific tumor suppressor genes. Diet, exercise, smoking and other environmental factors may have also contributed to their disease.

Human Genome Project

The Human Genome Project is an initiative of the U.S. Department of Energy that aims to generate a high-quality reference sequence for the entire human genome and identify all the human genes.The DOE and its predecessor agencies were assigned by the U.S. Congress to develop new energy resources and technologies and to pursue a deeper understanding of potential health and environmental risks posed by their production and use. In 1986, the DOE announced its Human Genome Initiative. Shortly thereafter, the DOE and National Institutes of Health developed a plan for a joint Human Genome Project, which officially began in 1990.The HGP was originally planned to last 15 years. However, rapid technological advances and worldwide participation accelerated the completion date to 2003. Already it has enabled gene hunters to pinpoint genes associated with more than 30 disorders.

Cloning

Cloning involves the removal of the nucleus from one cell and its placement in an unfertilized egg cell whose nucleus has either been deactivated or removed.

There are two types of cloning:

1. *Reproductive cloning*: After a few divisions, the egg cell is placed into a uterus where it is allowed to develop into a fetus that is genetically identical to the donor of the original nucleus.

2. *Therapeutic cloning*: The egg is placed into a Petri dish where it develops into embryonic stem cells, which have shown potentials for treating several ailments.

In February 1997, cloning became the focus of media attention when Ian Wilmut and his colleagues at the Roslin Institute announced the successful cloning of a sheep, named Dolly, from the mammary glands of an adult female. The cloning of Dolly made it apparent to many that the techniques used to produce her could someday be used to clone human beings. This stirred a lot of controversy because of its ethical implications.

BIOTECHNOLOGY IN AGRICULTURE

CROP YIELD

Using the techniques of modern biotechnology, one or two genes may be transferred to a highly developed crop variety to impart a new character that would increase its yield. However, while increases in crop yield are the most obvious applications of modern biotechnology in agriculture, it is also the most difficult one.

Current genetic engineering techniques work best for effects that are controlled by a single gene. Many of the genetic characteristics associated with yield are controlled by a large number of genes, each of which has a minimal effect on the overall yield. There is, therefore, much scientific work to be done in this area.

REDUCED VULNERABILITY OF CROPS TO ENVIRONMENTAL STRESSES

Crops containing genes that will enable them to withstand biotic and abiotic stresses may be developed. For example, drought and excessively salty soil are two important limiting factors in crop productivity. Biotechnologists are studying plants that can cope with these extreme conditions in the hope of finding the genes that enable them to do so and eventually transferring these genes to the more desirable crops.

One of the latest developments is the identification of a plant gene, At-DBF2, from Arabidopsis thaliana, a tiny weed that is often used for plant research because it is very easy to grow and its genetic code is well mapped out. When this gene was inserted into tomato and tobacco cells, the cells were able to withstand environmental stresses like salt, drought, cold and heat, far more than ordinary cells.If these preliminary results prove successful in larger trials, then At-DBF2 genes can help in engineering crops that can better withstand harsh environments. Researchers have also created transgenic rice plants that are resistant to rice yellow mottle virus. In Africa, this virus destroys majority of the rice crops and makes the surviving plants more susceptible to fungal infections.

INCREASED NUTRITIONAL QUALITIES

Proteins in foods may be modified to increase their nutritional qualities. Proteins in legumes and cereals may be transformed to provide the amino acids needed by human beings for a balanced diet. A good example is the work of Professors Ingo Potrykus and Peter Beyer in creating Golden rice.

IMPROVED TASTE, TEXTURE OR APPEARANCE OF FOOD

Modern biotechnology can be used to slow down the process of spoilage so that fruit can ripen longer on the plant and then be transported to the consumer with a still reasonable shelf life. This alters the taste, texture and appearance of the fruit. More importantly, it could expand the market for farmers in developing countries due to the reduction in spoilage.

However, there is sometimes a lack of understanding by researchers in developed countries about the actual needs of prospective beneficiaries in developing countries. For example, engineering soybeans to resist spoilage makes them less suitable for producing tempeh which is a significant source of protein that depends on fermentation. The use of modified soybeans results in a lumpy texture that is less palatable and less convenient when cooking.

The first genetically modified food product was a tomato which was transformed to delay its ripening. Researchers in Indonesia, Malaysia, Thailand, Philippines and Vietnam are currently working on delayed-ripening papaya in collaboration with the University of Nottingham and Zeneca.

Biotechnology in cheese production: enzymes produced by micro-organisms provide an alternative to animal rennet – a cheese coagulant – and an alternative supply for cheese makers. This also eliminates possible public concerns with animal-derived material, although there are currently no plans to develop synthetic milk, thus making this argument less compelling. Enzymes offer an animal-friendly alternative to animal rennet. While providing comparable quality, they are theoretically also less expensive.

About 85 million tons of wheat flour is used every year to bake bread. By adding an enzyme called maltogenic amylase to the flour, bread stays fresher longer. Assuming that 10–15% of bread is thrown away as stale, if it could be made to stay fresh another 5–7 days then perhaps 2 million tons of flour per year would be saved. Other enzymes can cause bread to expand to make a lighter loaf, or alter the loaf in a range of ways.

REDUCED DEPENDENCE ON FERTILIZERS, PESTICIDES AND OTHER AGROCHEMICALS

Most of the current commercial applications of modern biotechnology in agriculture are on reducing the dependence of farmers on agrochemicals. For example, Bacillus thuringiensis is a soil bacterium that produces a protein with insecticidal qualities. Traditionally, a fermentation process has been used to produce an insecticidal spray from these bacteria. In this form, the Bt toxin occurs as an inactive protoxin, which requires digestion by an insect to be effective.

There are several Bt toxins and each one is specific to certain target insects. Crop plants have now been engineered to contain and express the genes for Bt toxin, which they produce in its active form. When a susceptible insect ingests the transgenic crop cultivar expressing the Bt protein, it stops feeding and soon thereafter dies as a result of the Bt toxin binding to its gut wall. Bt corn is now commercially available in a number of countries to control corn borer, which is otherwise controlled by spraying.

Crops have also been genetically engineered to acquire tolerance to broad-spectrum herbicide. The lack of herbicides with broad-spectrum activity and no crop injury was a consistent limitation in crop weed management.

Multiple applications of numerous herbicides were routinely used to control a wide range of weed species detrimental to agronomic crops. Weed management tended to rely on Pre-emergence—that is, herbicide applications were sprayed in response to expected weed infestations rather than in response to actual weeds present. Mechanical cultivation and hand weeding were often necessary to control weeds not controlled by herbicide applications. The introduction of herbicide-tolerant crops has the potential of reducing the number of herbicide active ingredients used for weed management, reducing the number of herbicide applications made during a season, and increasing yield due to improved weed management and less crop injury.

Transgenic crops that express tolerance to glyphosate, glufosinate and bromoxynil have been developed. These herbicides can now be sprayed on transgenic crops without inflicting damage on the crops while killing nearby weeds.

From 1996 to 2001, herbicide tolerance was the most dominant trait introduced to commercially available transgenic crops, followed by insect resistance. In 2001, herbicide tolerance deployed in soybean, corn and cotton accounted for 77% of the 626,000 square kilometres planted to transgenic crops; Bt crops accounted for 15%; and "stacked genes" for herbicide tolerance and insect resistance used in both cotton and corn accounted for 8%.

PRODUCTION OF NOVEL SUBSTANCES IN CROP PLANTS

Biotechnology is being applied for novel uses other than food. For example, oilseed can be modified to produce fatty acids for detergents, substitute fuels and petrochemicals. Potatoes, tomatoes, rice tobacco, lettuce, safflowers, and other plants have been genetically engineered to produce insulin and certain vaccines. If future clinical trials prove successful, the advantages of edible vaccines would be enormous, especially for developing countries.The transgenic plants may be grown locally and cheaply. Homegrown vaccines would also avoid logistical and economic problems posed by having to transport traditional preparations over long distances and keeping them cold while in transit. And since they are edible, they will not need syringes, which are not only an additional expense in the traditional vaccine preparations but also a source of infections if contaminated. In the case of insulin grown in transgenic plants, it is well-established that the gastrointestinal system breaks the protein down therefore this could not currently be administered as an edible protein.

However, it might be produced at significantly lower cost than insulin produced in costly, bioreactors. For example, Calgary, Canada-based SemBioSys Genetics, Inc. reports that its safflower-produced insulin will reduce unit costs by over 25% or more and approximates a reduction in the capital costs associated with building a commercial-scale insulin manufacturing facility of over $100 million, compared to traditional biomanufacturing facilities.

CRITICISM

There is another side to the agricultural biotechnology issue. It includes increased herbicide usage and resultant herbicide resistance, "super weeds," residues on and in food crops, genetic contamination of non-GM crops which hurt organic and conventional farmers, damage for all wildlife from glyphosate, etc.

BIOLOGICAL ENGINEERING

Biological engineering is any type of engineering—for example, mechanical engineering—applied to living things. Examples: Aquaculture, Biosensors, Bio-based materials, Biomaterials, Industrial fermentation, Industrial enzymatic reactions, Production and purification of biopharmaceuticals, Advanced life support system, Artificial biospheres *e.g.*. Biosphere 2, Animal locomotion.

Bioengineers are concerned with the application of engineering sciences, methods, and techniques to problems in medicine and biology. Bioengineering encompasses two closely related fields of interest: the application of engineering sciences to understand how animals and plants function; and the application of engineering technologies to design and develop new devices, including diagnostic or therapeutic instrumentation, or the formulation of synthetic biomaterials, the design of artificial tissues and organs, and the development of new drug delivery systems.

Bioengineering is the application of the principles of engineering and natural sciences to tissues, cells and molecules. Closely related to this is biotechnology, which deals with the implementation of biological knowledge in industrial processes. Applications from both fields are widely used in medical and natural sciences and also in engineering.

WHAT IS BIOENGINEERING?

There are as many definitions of Bioengineering as there are groups working in the field. In the area of health, the US National Institute of Health formed a Bioengineering Definition Committee that released the following preamble and definition on July 24, 1997:

PREAMBLE

Bioengineering is rooted in physics, mathematics, chemistry, biology, and the life sciences. It is the application of a systematic, quantitative, and integrative way of thinking about and approaching the solutions of problems important to biology, medical research, clinical proactive, and population studies.

The NIH Bioengineering Consortium agreed on the following definition for bioengineering research on biology, medicine, behaviour, or health recognizing that no definition could completely eliminate overlap with other research disciplines or preclude variations in interpretation by different individuals and organizations.

DEFINITION

Bioengineering integrates physical, chemical, or mathematical sciences and engineering principles for the study of biology, medicine, behaviour, or health. It advances fundamental concepts, creates knowledge for the molecular to the organ systems levels, and develops innovative biologics, materials, processes, implants, devices, and informatics approaches for the prevention, diagnosis, and treatment of disease, for patient rehabilitation, and for improving health. If we ignore the obvious health focus in the NIH definition, it is clear that bioengineering is concerned with applying an engineering approach and an engineering focus to biological problems.

Bioengineering is the application of engineering design and technology to living systems. Any living system. While for the majority, the obvious, and perhaps justifiably, the most important, living system is the one walking upright and registered with a doctor, many bioengineers work in breweries, drug companies, farming and the environment. With plants, insects and fungi.

The discipline is thus not confined to the design and production of Medical Devices, but encompasses any situation where technology must interface with a living system. Bioengineers have to be mentally flexible, willing to adopt and adapt techniques from other industries, and to work with people from a wide range of disciplines.

As an example of the art, consider a Kidney Machine and the sub-systems which go to make up this example of vital life support equipment:

Water treatment and purification. Heating and temperature control for pasteurisation and patient comfort. Measurement systems for flow and pressure, electrolytes. Alarm systems for all vital parametres. Data collection and processing. Ergonomics Electrical Safety Clinical Measurement represents a large sector within the field. Modern sensor technology permits the measurement of parametres for aiding diagnosis and monitoring therapy to an extent impossible to imagine even ten years ago. Biomechanical engineers are designing tools enabling surgeons to perform sophisticated "keyhole" surgery, while device technology and computer complexity is advancing fast enough to bring the clinician's wildest aspirations within reach.

Bioengineering is what you do with biotechnology. Biotechnology is what you get from studying and learning how to manipulate biology. It has been known for more than a hundred years that life is basically a machine which takes in fuel and performs work. This machine is remarkable in that it arose through variation and selection from a very primitive self replicating molecule.

The closest analogue of this molecule present in our contemporary biology is the ribosome. It has been found that this ribosome consists, in part, of a strand of RNA. Inside the nucleus of a human cell there are twenty two pairs of chromosomes and either another pair or two different. These, as you are probably tired of hearing, are mostly DNA. In our cells is also an organelle which is called a mitochondria.

This organelle was, once, a separate life-form. It evolved separately from the type of cells that make up most of our body and its internal structure is significantly different from the cell it is now a part of. These mitochondria have their own DNA but have come to live within our cells as symbiotes. . What is the significance of DNA? It's just a molecule, right? Yes it is. That is exactly what it is. It has no magical properties whatsoever. Science fiction notions of being able to mutate an adult human by changing their DNA to that of another species, perhaps, would not have any morphological change on the person because there is not, yet at least, a gene that can reshape bone. Most random gene changes result in some kind of disease such as cancer. This should seem weird to you.

I am talking about changing a DNA molecule but I'm not suggesting that it becomes anything other than a DNA molecule. I mean changing your garden variety molecule by changing even a single bond usually makes something totally different. What is DNA that it is different? For one thing DNA is actually a family of molecules composed of adenine, thymine, cystine, and guinine. RNA uses a molecule called uracil in place of thymine. A strand of DNA is a linked chain of any number of any combination of those four "base pairs". When we talk about modifying DNA we are talking about rearranging sequences of base pairs.

The DNA molecule serves a function no different from the hard drive on your computer. It stores information. The three billion year history of the DNA molecule is a testament to its effectiveness as a storage medium. RNA is more like your computer's working memory. It is natural to consider the act of designing a DNA strand in the same way we would consider the act of writing a programme for a computer.A bioengineer, or as it was put in Blade Runner, a genetic designer, is a DNA programmer. I am trying to express these ideas in these terms because a certain mysticism has arisen around DNA that needs to be dispelled before we can consider the questions of what we might be able, or want, to do with our newfound ability to manipulate it.

There are a great many who are so shocked by these ideas that they immediately rush to outlaw research as soon as they are informed of its possibility. These feelings of fear and anxiety extend to many areas of reproductive medicine and general angst about self-modification. All these things boil down to one simple fact. People are afraid to acknowledge what they are. If they ever dared, they may have tried to feel their bodies as being made of cells and molecules and existing in a void of randomly moving gasses and a hard empty floor.

They might be able to open their eyes wide enough to see the functioning of their minds and understand that it is merely a computation running on some very impressive but ultimately kludged together hardware. I do this from time to time to keep my bearings. It is not a pleasant experience. Throughout history this truth has been an inevitability. No matter how keen your vision is there was nothing you could do about it. Being, quite literally, helpless people constructed fictions about souls, gods, and the tooth fairy and just about any other entity or idea that people tend to bring into these debates. These are beloved lies.

Until now, they've been the lies that have allowed us to live. With bioengineering as suggested, have the power to change that. As suggested, soon have complete power. You no longer have to be afraid of the beat of your heart knowing that each one has a number and the number is counting downward. You no longer have to feel alone in your own skull. You can have a neural interface.

You will no longer need to fear crippling injury, it can be repaired. You will be able to live for years in zero gravity in perfect health. You will be able to tolerate significant doses of radiation. You will be able to metabolize poisonous chemicals. You will never suffer a serious illness. These are the possibilities.

All we, as a people, need to do is face our fears with the sword called Biotechnology. Lies need no longer be our only shield. In the past all changes to your body were either impossible or irreversible. To live a long and healthy life one had no choice but to take the best care possible of their bodies so that it would remain healthy for as long as possible. For this reason the body became an object of worship. Something that deserves special consideration above all others.

While the health and safety of our embodiment will always be a concern it is no longer necessary to treat it as holy and untouchable. It is foreseeable that people might be struck by some fancy or other that they want to follow for a few years. These people will no longer be forced to consider their long term health because it will be taken care of by advanced medical nanotechnology which will be able to implement or reverse just about any change.

Considering that there is nothing in the universe that has any opinion one way or another what we do to ourselves we might just as well do whatever we want. All of you have are your desires and your tools. It would be a tragedy in more ways than one if the conservative elements in our psyches strangle this one eighty year opportunity we are given to do what ever we want with our body. There is no such thing as a crime against one's self and nothing should be considered as such. We live in a free universe if not in a free country. So to these conservative voices I ask "why not?"

A LIVING FUTURE WITH ROOM FOR EVERYONE

On the other extreme there is a faction in the singularitan community which thinks that people who wish to use biotechnology are, to pick a word, backwards. They think that the singularity means that their nanoscale robots are inherently better than the biological cells which have served us from, practically, the beginning of time merely because they happen to process information in a slightly different way which would allow them to be faster and more vicious in the war of evolution.

Although their machines are made out of the same atoms that compose the bulk of our own bodies, they have told me that their "machine phase life" will quickly supersede biological life. They wish to plunge the universe into a new epoch of life which they will direct from their own computer simulations as uploads. To these people my question is WHY?

Creating a "doom goo" is behaviour fit for a script kiddie, the lowest grade of computer abusers. It would be no different from the Mac users getting their final revenge on all the PC users. Civilized people who choose to use ultra-technologies should have the, well, civility to respect the people who want to take a little more time to figure out what they want to be or to be something completely different.It is wrong, in every way, to unleash such destruction on people who only wish to remain human regardless how superior the replacements are. The only reason we develop technology at all is to make our own, human, lives better. Lets take these nanotechnologies and let them become our symbiotes just as the mitochondria are.

Lets build a massively powerful computer not to run simulations of our brains but rather simulations of our proteins. Let us apply bioengineering to our own cells to make the very stuff of our body better than it has ever been before.The ultimate use of nanotechnology is in our own cells. We, ourselves, in our present incarnation can take advantage of everything I am talking about here. As suggested, start with the cell and re-engineer it from the first atom. As suggested, use nanites, DNA, hydrocarbons, or whatever else to make a stable, resilient, and vital whole.

Our new biologies will be limited only by our imaginations and the laws of physics. These bold dreams excite only ridicule in the most radical computronium-heads. The universe is billions of light-years across and they don't see any room in it for us.

This is one of the most important areas that this force should be applied. The singularity is a neutral thing. It can be spectacularly good if it can be kept under human[oid] control. It has an equal potential to be unimaginably bad for everything we, today, call living things. As radical and audacious as my ideas about cyborgs and bioengineering may seem they are, in the grand scheme of things, quite conservative and an appropriate mean between the two extremes. Lets bring on the future but lets do it with our eyes open and our own interests in heart and mind.

SCIENCE, TECHNOLOGY, AND THE ART OF DESIGN

Body blueprints are passed around all the time by their own natural functions. Only in the last fifty years we have been able to appreciate sexual reproduction as an information exchange. In the muck at the bottom of ponds you can find that the bacteria that live there have established a commune of sorts in which DNA is passed back and shared among them.

Things won't be so easy-going with the blueprints of the new cellular structure I mentioned because of the intense effort required to produce them. I think that because of this and the importance of a fair and equal access that the core sciences and technologies behind cyborg biologies, and everything else for that matter, should forever be kept in the public domain.

The people who think it would be only fair that the people who put the direct effort into the direct design work should receive a patent for a few years in order to recoup their costs. Lets think about this. How did these people get the scientific knowledge about DNA and the proteins? Where did they get their computers?

Who provided the computing services? How did they figure out how to programme the AI that helped them organize the gigabytes of data that is required to build a cyborg and its accompanying technologies? The answer to all of these is millions of people. More than a million biologists and chemists made significant contributions to the project.

Every single person who has purchased a computer has helped fund further research and industrial development that made the project possible. Hundreds of thousands of people contribute their spare computer cycles to on-line projects such as folding@home which studies the very proteins that are the subject.

Their AI wasn't the result of some eureka moment but the culmination of five thousand years of mathematical research. They also owe debts to the neurologists and psychologists who have spent the last two hundred years mapping the brain.To take all this effort from practically the entire human race and then try to sell it back to us at a price is outrageous. DNA and any nanotech design is merely data which, in fact, can be stored on a hard disk. This means that you can download it to your hard disk from the internet. As the science of cellular biology is completed the finished proteins and core science should be made available on the internet to anyone who wants it. This might not seem very useful to you as an individual. I will return to that shortly.

This isn't to say that all data should be made available. There is a difference between the right to know everything about a brick and the right to know how to build a luxury home. Unlike the basic science and technology which is our birthright the actual integrated designs of original bodies are a form of art. The service of designing a body is one that should be paid for. Designers work alone from the common knowledge and create something new that is from themselves and hence worth being sold.

More generally it is best that designs be kept private so that there will, necessarily, be many more of them and greater variation. It is not at all my goal for anyone else to even be the same species as me except, perhaps, my offspring. Others might say that it is always best to be an upload running on computronium. Some say that doing anything at all to one's self is wrong. Aside from my own personal desires, my goal is to defend everyone's right to be what they want to be. I want the singularity to be an enabler of deep personal desires rather than the realization of some grand master plan. The latter would be an absolute horror.

One of the concerns that might be raised regarding the notion of keeping designs proprietary is how all these, potentially radically, different breeds of people will interact in society. A major concern will be whether these new species will be able to form families and have children like humans do. This is not a terrible concern as there will be the possibility of creating offspring from scratch. This may not be the most reliable means as one would prefer not to abandon the old way as a fallback.

With intelligent machines it should be possible to take even the most different designs and to generate some integration of the two. This function can be integrated into the body of a cyborg female or hermaphrodite. With these two functions of differentiation through separate design and intelligently controlled recombination the mechanism of evolution will continue into the future but not in its old merciless clothes.

HYPER-EVOLUTION

The process of evolution is about to be changed radically. In the past evolution proceeded on the basis of genes that are passed from parent to offspring and changes that arise from semi random mutations. In the future traits will be passed in the form of binary strings and edited by the intelligent design of conscious minds. Ever since we gained the ability to talk our minds have had the ability to create, transmit, and process things that are sometimes called memes.

A meme shares many of the same properties as genes. The period of time between the creation of language some 40,000 years ago and these decades in which we live today when the genome was cracked and will soon be the subject of arbitrary manipulation can be understood as a process by which memes have arisen from genes and now are in a position to replace them. This replaces the evolution of genes with what I call the hyper-evolution of memes. I mean hyper-evolution to mean nothing more or less than evolution by conscious design. In terms of biology alone the coming years promise to be very chaotic.

The new reflexive relationship of an individual to his own genetic makeup will result in a dynamic feedback loop which will significantly alter the historic patterns and equilibrium. It is reasonable to assume that after a few decades or centuries the process will result in some new stable equilibrium. It is very difficult to speculate on the dynamics of such a process as there is little historic data to fall back upon.

The closest analogue might be the evolution of the automobile or of computer software especially open source software. Hyper-evolution is very good news for individuals because it disconnects the process of evolution from the lives and deaths of individual bodies. As Dawkins said, a chicken is an egg's way of creating another egg. With nano-enhanced cells we can disconnect our lives from the process which selects our traits.

The classical idea of eugenics can be completely discarded. It is no longer necessary to make sure that people have the right genes at birth because they can be changed later in life. The easiest way to make such a change would be to add the nano symbiote to the zygote. These nanites could be activated at any time in the individual's life to whatever effect desired. Retrofitting an adult individual with these will be immensely more difficult but not altogether inconceivable.

All evolution is renown for its cruelty. It "de-selects" traits by killing individuals. Proposals for directed evolution from Plato forward have been little better. Because hyper-evolution has the potential to select and propagate traits without the necessity of creating and destroying individuals.

This means that the way we contemplate the population dynamics of a hyper-evolutionary system vary differently from the traditional. Instead of considering the population of chickens in the universe we consider the populations of memes in the hosts that are capable of sustaining them.The fitness of a given meme is therefore dictated by the probability of it being chosen by a given host and by how well it serves that host in order that it may be propagated. In this way the memes that we can transmit through computer networks have the relationship to us as DNA does to a cellular membrane which houses it. To us, it means that the race of competition between individuals will be over as the war of replicating memes has moved to a new level. We have the prospect of an eternity of peace, health, opportunity, and unlimited choice. A threat to this is the observation that evolutionary forces might interact with government structures as a means to self-propagation. The value of personal freedom must be maintained all others.

You can be bio-engineered to be, practically, immortal. Age will mean nothing to you and accidental death won't be nearly as great a risk as it is today as your body will be more resistant and resilient to all but the most grave injuries. The dynamics of evolution have made genetics extremely good at constructing bodies. Unfortunately there was no selective pressure pushing for better mechanisms of maintenance and repair. In the future not only the more ambitious among us but also the very memes that will specify the fabric of our bodies will select for the most stable possible host. You may not choose to take advantage of this for yourself but some will.

There is nothing magical about the elixir of life that's coming to a health food store near you. It will merely be a type of nanotech which will be designed to repair and, occasionally, tweak the designs of your cells. In this manner your body can be perpetuated indefinitely. There are no catches, no hidden costs, you won't even have to sell your soul. You will, if you choose to, simply continue to live as long as you like. You will not be constrained to a decrepit old body either. Your nanites can restore you to any appearance or level of health you desire. A few years after this is brought to market the cost will come down so low that your next body could be cheaper than the car you might have bought last month.

Some people object to immortality based on the idea that they should make room for the next generation. That's not a problem because soon as suggested, be able to emigrate to other planets. You could become a homesteader on Mars. The question of whether some people should seek immortality is not an issue because it would be highly unethical to kill someone who wants to live. The problem arises in the cultural conflicts between the community of immortalists and the traditionalists. The key is tolerance on both sides.

Each person must make their own choice about whether to try for immortality. There shouldn't even be a universal code of all bodies must be immortal unless otherwise specified. As long as the choice remains open there is no problem. We can think of lives like rockets.

Every once and a while we design one, haul it out to the pad and hit the big red button. In many cases these rockets have exploded into flames, never getting off the launch pad. Others launch, make it all the way to the edge of space, and do whatever small mission is theirs and come down without ever reaching orbit or orbiting for only a few years before coming down.

Sometimes they go up there, serve out their useful life and then continue to operate for many decades, sometimes exceeding their intended lifespans by a large multiple. In 1974 the Amateur Satellite service launched the seventh in their Oscar series. It continued to work over the next seven years until the battery shorted out. Sometime in the two decades that followed the short was cleared by the constant flow of electricity. Today it only needs the sun. Its beacon is still loud and its transponders open and clear after having gone around the planet some 127,000 times.

Likewise some of us will live to see the year 3,000 and some won't. There is no imperative to change the natures of the people who simply don't want to live so long just as there is no need to set some expiration date for those who do. Nobody has yet lived past 135 or so. Reports of longer lifespans are dubious.We can't say we know for sure what it really means to live so many years. We can say with confidence that we can make a humanoid body that doesn't grow old. We cannot say what will happen to the mind over a long period of time. Studies have shown that the brain looses a significant amount of plasticity during adolescence and the slowness of the elderly to adopt new ideas is almost a cliché. If you think about it too hard you start questioning the meaning of life and all the other fundamental questions of existence. In the future these questions will no longer be abstract philosophy but matters of practical importance.

Let's say that you have decided to choose immortality and you are now 120 years old. The world is changing and it is becoming clear, even to you, that you will need to adapt to keep up. In theory it will be possible to use neural stem cells to restore the plasticity of youth.

When you do this you discover that while this does, indeed, restore your ability to learn new skills and ways of thinking it overwrites parts of your mind so that your memories of the past tend to be less clear and your personality starts to shift towards the modern norm.

You feel that you have become a new person through biotechnical intervention. While your body and mind remain continuous your inner self has been changed in such a way that it seems, to everybody, that the old you is dead.Another approach to solving this problem would be to gradually replace your brain, as it deteriorates, with a physical implementation of the best AI around. This new brain works very differently from your original. You chose it because it also performs about a million times better in every way and offers a number of interesting extended capabilities that lie outside of the scope of this essay.

While such a brain would be able to function at its peak indefinitely into the future there is a serious question of whether your mind can do like a hermit crab and move into the new host.It can be foreseen that in the protracted future people who choose not to upgrade their original brains significantly will be at a serious disadvantage.

Only when we understand the mind well enough to construct artificial intelligences will we be able to begin to answer these questions.

AN INVITATION TO DREAM AND LIVE

The bulk of this essay has been about the basic concepts and issues surrounding bioengineering. I could just tie it up with a nice cease. If I did that I would give up the chance to share some of the hope I have for this coming age. Bioengineering didn't just arise out of many diverse human activities it can also serve people in ways hardly even imagined now.There is no reason to be ashamed of wanting to change, even if it is into something very different. It is a wonderful thing to have desires, even crazy ones. Bioengineering will enable us to go places that our current bodies can't even survive such as the oceans or worlds that receive too much radiation for our current forms. Even if your interests are much more personal bioengineering is still the right tool.

It is time to begin dreaming like children again because it is our dreams that are the fuel of all adventures. The times ahead will be both strange and wonderful. With hard work and constant vigilance far more dreams than nightmares will come true. The singularity itself is a nexus of dreams and terrors. To make it happen for us we must invest our dreams in it with the knowledge that our hard work will bring them true. We can make it wonderful for us and everyone.

If, perchance you are not yet inspired by my presentation or simply don't want to change your body in any significant way I have perfect sympathy for you and will respect you always. If you think that there is something inherently wrong with what I am talking about here and think that these developments must be halted I ask you to reconsider. Without considering any modification to yourself, think of all the costs of maintaining your current body.

Consider the prescriptions, the doctor's appointments, and the costs of a visit to a hospital should you run into some misfortune. Now think about the most advanced hospital in the world staffed with the best doctors in the world. Think of a hospital which can provide you with every treatment or medicine you could ever need or want. Now imagine that it only cost you five dollars per visit. This is not a joke, this is possible.

Consider the technologies of AI, Nanotechnology, and Biotechnology. Now imagine all these wonders integrated into a single pod which you could purchase for as little as $2,000. Medicine can be automated, completely. You can purchace one of these for yourself or your family and only need to pay for the raw molecules required to construct the supplies it needs.This pod will contain within it a mind a hundreds times smarter than the most brilliant physician that ever lived and completely devoted to your service. Its internal systems will be able to integrate new procedures and inventions simply by downloading the blueprints from the internet and performing an auto-upgrade. You will be free of the medical establishment entirely. Such pods should only be regulated to the minimal extent required to ensure the accountability of their designers for any malfunction.

WHAT DO BIOENGINEERS DO?

Bioengineers deal with a wide variety of problems. Students with bioengineering degrees may work as biomedical engineers with medical practicioners to develop new medical techniques, medical devices, and instrumentation for manufacturing companies.Clinical engineers work in hospitals and clinics to maintain and improve the vast amount of technological support required in modern medicine. With advanced degrees in the various fields of bioengineering, some graduates perform basic research related to biology and medicine in the research laboratories of educational and governmental institutions or in the medical industries.

Bioengineering is the application of engineering techniques and understanding to medical or biological systems. Traditionally medical research has been performed by medical doctors. It was found, however, that sometimes Doctors ended up re-inventing or re-discovering things that had already been understood or discovered in the physical sciences because the Doctors were unaware of what had been studied in those sciences. Similarly when engineers turned their attention to medical problems they often found that the problems that were interesting from a physical science perspective were unimportant from a medical perspective.

Recently, therefore, there has been a huge increase in collaboration between Medical Doctors and Engineers. Examples of problems that have benefited from this type of collaboration are the development of artificial organs, laser and radiation treatments for cancer, and improved imaging technology such as the MRI.

Collaboration has also begun to occur more often between engineers and people in other biological sciences such as marine biology. In these areas, once again, engineering techniques are being applied to biological systems to provide increased insight and better models of the biological world. At UMass Dartmouth one such collaboration is the modeling of plankton transport in turbulent ocean flow.

Mechanical engineering plays a large role in bioengineering because biological systems quite often involve fluid flow, such as blood flow, ocean currents and wind, as well as heat effects, transport of chemicals and mechanical stresses. These days almost every mechanical engineering department has at least some faculty working in the area of bioengineering. Aquaculture is the cultivation of aquatic organisms, such as fish, shellfish, algae and other aquatic plants. Mariculture is specifically marine aquaculture, and thus is a subset of aquaculture. Some examples of aquaculture include raising catfish and tilapia in freshwater ponds, growing cultured pearls, and farming salmon in net-pens set out in a bay.

Aquaculture has been one of the fastest growing segments of global food production in recent decades, and has been hailed as an answer to declining wild fish stocks caused largely by overfishing.

Tuna farming in Australia, as well as of other species, has had immense success. In some limited or specialized locations, United States, Hawaii aquaculture embraces a wide range of experimentation NELHA with applications for various combinations of DOW Deep Ocean Water, SOW Surface Ocean Water and renewable energy sources Gateway with commercially available electricity and fresh water.

Challenges In countries like the U.K., Canada, Norway, and Chile, salmon and trout farming are one of the fastest-growing forms of agriculture. Salmon farming, like other food producing operations such as beef, wheat or tomatoes can impact the environment. In particular organic wastes from fish cages can have a significant effect on water quality and the population structure of organisms, increasing the occurance of toxic algal blooms, as has been the case in Scotland, but even a month of fallow time can return the area to pristine condition.

Like other agriculture production, it must stand up to a rigourous evaluation of any environmental impact. Salmon aquaculture has come under increasing scrutiny from environmental Non-governmental organizations. In Canada, salmon farming sites occupy a small portion of the coastal zone areas where they are located. The total area occupied by Canadian salmon farms in British Columbia and the Bay of Fundy in New Brunswick is about 8,900 acreas which is less than 0.01% of the coastal area where these sites are located. Bioengineering involves the study of the human body from the engineering perspective. It is an emerging field which combines several disciples including mechanical, chemical, and electrical engineering with human and animal physiology to develop engineering solutions to the body.

In college, students will take engineering courses along with biology and anatomy courses to learn about the human body. Bioengineering may be offered with a pre-med programme. If you are interested in both engineering and medicine, this field should be considered. Examples of bioengineering applications in medicine include the pacemaker for heart patients, artificial limbs and joints. Bioengineering concentrates on combining biological science and engineering to study, define and utilize the interaction of these disciplines to solve engineering problems. It focuses on the utilization of plants and animals for the benefit of humans and to maintain a healthy environment. a, f. Students in the area need a strong background in biology, particularly at the cellular and organism level, chemistry, fluids, processing, thermodynamics, and physiology. Bioengineering is the use of engineering principles of analysis and design to solve problems in medicine and biology.

Wild Pacific and Atlantic salmon stocks have seen significant declines over the last several decades, before salmon farming operations started. These declines were caused by a combination of factors including climate change, overfishing and freshwater habitat destruction. Canadian salmon farmers have significantly reduced the escape of their salmon. The evidence shows that the escape of farmed salmon on Canada's west coast poses low risk to Pacific salmon, however concerns have been raised on the East coast that wild Atlantic salmon may interbreed with salmon that escape from farms.

Many farmed fish species are carnivorous, meaning that other wild fish species must be harvested to maintain the fish farm, but these are species which are not used for human consumption. A large portion of the fish meal used in fish feeds comes from the trimmings and discards of commercial species. More and more feeds are made using poultry and vegetable oils as substitutes for fish oil.

Other problems with aquaculture include the potential for increasing the spread of unwanted invasive species, as farmed species are often not native to the area in which they are farmed. When these species escape, they can compete with native species and damage ecosystems. Another problem is the spread of introduced parasites, pests, and diseases.

Mariculture is the cultivation of marine organisms for food, either in their "natural environment" or in seawater in ponds or raceways. An example of the latter is the farming of marine fish, prawns, or oysters in salt water ponds. By definition, mariculture is a specialized branch of aquaculture.

The Japanese have developed a clever process for free ranching marine fishes. The principle is based on behavioural conditioning and the migratory nature of certain species of marine fishes. The fishermen first raise fish hatchlings in a closely knitted net in a Harbour. They sound a underwater honk before each feeding.

When the young fishes are old enough, the fishes are freed from the net. The fishes grow up in the open sea. During spawning season, these fishes return to their birthplace. The fishermen harvest the fishes by sounding the honk and then raise the net.

Mariculture is not limited to food production only, products such as cultured pearls are considered mariculture as well.Off the coast of California, the top few feet of natural kelp beds are harvested by boats with mowers. Kelp provides alginin, an edible material used in ice cream and cosmetics.

A biosensor is a device for the detection of an analyte that combines a biological component with a physicochemical detector component.

It consists of 3 parts:

1. The sensitive biological element can be created by biological engineering the transducer in between the detector element. Arrays of many different detector molecules have been applied in so called electronic nose devices, where the pattern of response from the detectors is used to fingerprint a substance.
2. A canary in a cage, as used by miners to warn of gas could be considered a biosensor. Many of today biosensor applications are similar, in that they use organisms which respond to toxins at a much lower level than us to warn us of their presence. Such devices can be used both in environmental monitoring and in water treatment facilities.
3. A bio-based material is simply an engineering material made from substances derived from living tissues. These materials are sometimes referred to as biomaterials, but this word also has another meaning.

Strictly the definition could include many common materials such as wood and leather, but it typically refers to modern materials that have undergone more extensive processing. Milk is an excellent food source for humans and bacteria alike. It is full of vitamins, fats, minerals, nutrients and carbohydrates. It is rich in the protein casein which gives milk its characteristic white colour. The most abundant carbohydrate is the disaccharide lactose, "milk sugar." At room temperature, milk undergoes natural souring caused by lactic acid produced from fermentation of lactose by fermentative lactic acid bacteria.

This accumulation of acid decreases the pH of the milk and cause the casein to coagulate and curdle into curds and whey. Curds are large, white clumps of casein and other proteins. Whey is the yellow liquid that is left behind after the casein has formed curds. Thus, bacteria obtain nutrients from the milk, inadvertently curdle it and humans use it as the first step in making many dairy products. The microbes important for dairy product manufacturing can be divided into two groups, primary and secondary microflora.

Products undergoing fermentation by only primary microflora are called unripened and those processed by both primary and secondary microflora are called ripened. Primary microflora are fermentative lactic acid bacteria which cause the milk to curdle. e, j, c. During dairy product production, milk is first pasteurized to kill bacteria that cause unwanted spoilage of the milk and of the downstream milk products. Primary microflora consists of certain kinds of Lactococcus, Lactobacillus and Streptococcus that are intentionally added to pasteurized milk and grown at 30°C or 37°C.Secondary microflora include several different types of bacteria, yeasts and molds; they are only used for some types of surface ripened and mold ripened cheeses. The various combinations of microflora determine what milk product you will end up with.

BIOTECHNOLOGY IN ANIMAL AGRICULTURE

Since the onset of the modern era of biotechnology, scientists have made impressive strides in developing new agricultural biotechnologies. Biotechnologies that enhance productivity and productive efficiency I have been developed and approved for commercial use. Technologies that improve productive efficiency will benefit both producers and consumers because feed constitutes a major component of farm expenditures.

New approaches by which animals may be a source of both biopharmaceuticals for human medicine and organs for transplant also are under development. Advances in biotechnology research have allowed impressive improvements to be made in diagnostic approaches, increasing microbial safety of food and improving animal health. The application of genomics and bioinformatics in animal agriculture will provide new genetic markers for improved selection of all livestock species.

Transgenic biology provides a means of altering the genome of animals to achieve desired production or health outcomes of commercial value and societal importance. Biotechnology also offers considerable potential for animal agriculture as a means to decrease the nutrients and odors from manure as well as the volume of manure produced. Development and adoption of these biotechnologies will contribute to a more sustainable environment.

The discovery and development of new animal biotechnologies are part of a continuum leading to the commercialization of agricultural biotechnology products. To enter the marketplace, new products of agricultural biotechnology are evaluated rigorously by the appropriate federal regulatory agencies to ensure efficacy, consumer safety, and animal health and well-being.

To benefit agriculture and society, products of biotechnology must be accepted by consumers. Central to consumer acceptance is the need to provide effective population-based education programmes to enhance public understanding of the safety and benefits associated with technological advances enabled by agricultural biotechnology.

The objective of this paper is to summarize a variety of existing and emerging biotechnologies and their current or potential benefits to societies around the globe. This overview is the first in a series of brief papers introducing several topics related to animal biotechnology.

HUMAN HEALTH: USES OF BIOTECHNOLOGY IN ANIMAL AGRICULTURE TO DECREASE MORBIDITY AND MORTALITY FROM DISEASE

Livestock have long been used to produce medicines for humans. Before the advent in 1982 of recombinant deoxyribonucleic acid technology for manufacturing insulin, the insulin needed by diabetics was extracted from animal tissues obtained as a by-product of meat production. Animals also have been the source of products such as heparin, heart valves, various sera and antisera, and collagen.

Advances in recombinant DNA technology, animal embryology, immunology, and other disciplines give rise to the prospect that animals will become important sources of highly sophisticated biopharmaceuticals and biological products. A number of biopharmaceutical products are being developed in which the ultimate production system will be a genetically modified animal. In these instances, the gene for the desired molecule is designed and constructed so it can be expressed only in a specific tissue. Several companies focusing on these activities are targeting the production of specific animal proteins in milk and in eggs.

These new approaches will allow for more economical production of certain pharmaceuticals that currently are expensive to produce. Biotechnological production methods thus may allow for more widespread distribution and application of pharmaceuticals worldwide.

Another type of biotechnology in development involves using animals to produce donor organs for human transplant. Heart valves from pigs have been used for many years to replace heart valves in humans, and several private-sector companies and numerous university scientists are investigating ways to use biotechnology to develop pigs as a source of transplant organs. One key advantage of this approach is the possibility that genetic engineering can be used to produce immune, rejection-free organs. In each instance just described, a vital basic health need will be fulfilled by the use of animals, and no foreseeable alternatives exist.

FOOD PRODUCTION: USES OF BIOTECHNOLOGY IN ANIMAL PRODUCTION

Feeding Livestock

Biotechnology has led to the development of plant varieties with improved qualities including enhanced tolerance of herbicides and protection against viruses and insect pests. These varieties have been adopted rapidly by American farmers, and the United States accounts for approximately 30% of the global area of transgenic crops. Genetically modified crops used for livestock feed include corn, soybean, canola, and cotton. In the United States approximately 80% of the corn and 70% of the soybeans are used for livestock feed.

Health and safety are priorities in the development of new food and feed products, including those developed through biotechnological, or "biotech," means. Evaluation by governmental regulatory agencies is required for each new biotech plant used for feed or food. Scientific studies evaluating feed components derived from GM plants have focused on beef cattle, swine, sheep, fish, lactating dairy cows, and broiler and layer chickens, and have included nutrient composition assessments, digestibility determinations, and animal performance measurements. Evaluation of nutritional composition has included micronutrients as well as macronutrients such as the content of specific amino acids and fatty acids. Likewise, digestibilities of nutrients derived from GM plants have been compared with feed components from near-isogenic control plants and Non-genetically enhanced feed sources.

Finally, feedstuffs derived from GM plants have been evaluated in performance studies conducted during growth phase and lactation. Performance results provide very sensitive indications of animal well-being and capture metabolic aspects in the use of absorbed nutrients as well as effects on the maintenance of normal physiologic functions. Evaluations have shown uniformly that feed components derived from GM plants commercialized thus far are substantively equivalent in terms of nutrient composition and are similar in terms of nutrient digestibility and feeding value.

Overall, feed components of GM plants result in growth rates and milk yields not different from those derived from Non-genetically enhanced feed sources. Studies have reported that when corn has been altered genetically for protection against the corn borer, under certain growing conditions GM plants can have lower mycotoxin contamination, resulting in safer feed for livestock.

In addition to evaluating the direct effects of feed components derived from GM plants on livestock performance and well-being, research has been conducted to assess the indirect effects involving the digestive fate of the DNA and protein introduced in GM crops, the potential transfer of DNA and protein, and their accumulation in food products such as milk, meat, and eggs.

All foods and feeds contain a variety of DNA and proteins from many sources, including plants and animal products. Protein and DNA in feeds and foods, whether from GM crops or Non-genetically enhanced sources, typically are degraded during normal digestion. All scientific evidence indicates that DNA introduced in transgenic plants and the proteins they encode are degraded during the normal digestive processes, just as DNA and proteins from Non-transgenic plants are degraded. There has been no detection of transgenic DNA or proteins in foods derived from animals that consume feed components from GM plants.

Metabolic Modifiers

Advances in understanding the regulation of nutrient use in agricultural animals have led to the development of technologies referred to as metabolic modifiers. Metabolic modifiers are a group of compounds that modify animal metabolism in specific and directed ways. They have the overall effect of improving productive efficiency, improving carcass composition in growing animals, increasing milk yield in lactating animals, and decreasing animal waste per production unit.Two classes of compounds have received major focus:

somatotropins and ß-adrenergic agonists. Somatotropin is a protein produced by the pituitary gland that differs slightly in structure among animal species. Thus, commercial application of STs depends on the use of recombinant DNA technology to produce the ST protein specific for a species. ßadrenergic agonists represent a class of compounds called phenethanolamines, and individual compounds differ in their biological effect. The several that affect animal growth are often referred to as repartitioning agents. Metabolic modifiers are considered to be new animal drugs and are regulated as such by the Food and Drug Administration's Centre for Veterinary Medicine.

The most commonly discussed ST is bovine somatotropin, which has been administered to dairy cows to achieve increased milk yield, unprecedented improvements in productive efficiency, and decreased animal waste. Specific biological mechanisms for the effects that bST has on nutrient-use efficiency and animal wellbeing have been identified. Commercial sales of bST began in 1994, and use gradually has increased so that approximately half of U.S. dairy herds are receiving bST supplements.

This usage involves herds of all sizes representing a full range of management systems from all U.S. geographic regions. Commercial experiences have demonstrated further that bST supplements result in marked improvements in productive efficiency while maintaining normal cow health and herd life. Bovine somatotropin is being used commercially in 19 countries.

Experimental studies of ST use in growing animals have focused on porcine somatotropin. Administration of pST to growing pigs results in greater nutrient use for lean tissue and less for body fat.This shift in nutrient partitioning results in substantially improved feed use, and the shift in lean:fat ratio represents an unprecedented improvement in carcass quality. Biological mechanisms that account for the effects of pST have been delineated and involve coordinated changes in lipid, protein, and carbohydrate metabolism. Improvements in dietary protein use efficiency with pST are especially significant and represent a decrease in the nutrient requirement per unit of lean tissue gain. Porcine somatotropin is undergoing testing required for FDA approval. Worldwide, pST is approved for commercial use in 14 countries.

Supplements of ß-adrenergic agonists to growing animals improve feed use and increase the rate of weight gain, carcass leanness, and dressing percentage. Researchers have established that the mode of action involves changes in endocrine and cellular mechanisms.The net effect is that these repartitioning agents improve productive efficiency by modifying specific metabolic signals in a coordinated manner to increase nutrient use for lean tissue accretion.Ractopamine is the only ß-adrenergic agonist currently approved in the United States—in this instance, for finishing pigs. Its commercial use began in 2000. Worldwide, ß-adrenergic agonists are approved for commercial use in 11 countries.

Transgenic Animals

The ability to alter the genome of animals by introducing DNA is a major technological advance in biotechnology and animal agriculture. Transgenic animals are produced by the introduction of a small, isolated, known fragment of DNA into preimplantation embryos. This DNA is inserted into chromosomes of the embryo and is expressed in tissues of the resulting individual. The ability to move genes into organisms also is referred to as gene transfer and is a technique of great importance not only to agriculture but to medicine as well.

Production of transgenic livestock provides a method of rapidly introducing "new" genes into poultry, cattle, swine, sheep, goats, and fish without crossbreeding, transporting live animals, or using semen, all of which can be potential means for disease transmission. There are numerous potential applications of transgenic methodology to develop new or altered strains of agriculturally important livestock.

Practical applications of transgenic biology in livestock production include improved milk production and composition, increased growth rate, improved feed use and carcass composition, increased disease resistance, enhanced reproductive performance, increased prolificacy, and altered cell and tissue characteristics for biomedical research and manufacturing. Development of transgenic farm animals will allow for more flexibility in the direct genetic manipulation of livestock These modifications are regulated as new animal drugs by the FDA's Centre for Veterinary Medicine.

The overall goals of producing transgenic animals are to enhance the knowledge of biology and biomedicine and to increase the efficiency of milk, meat, and fibre production. To realise these goals, desirable traits must be quantified and the genes responsible for them identified and introduced into production livestock. A process must occur of "select and/or re-design" for populations of superior individuals that can be propagated. It is important to note, however, that the production capability of genetically selected and/or genetically engineered animals will be realised only when their true genetic potential is attained through appropriate environmental and management considerations.The ultimate utility and value of transgenic technology will be determined by the ability to identify genes for the production of traits desired for improving transgenic animals as well as the ability to incorporate these desired genes, in an appropriately regulated manner, into domestic livestock. The rate at which genetic improvement technologies are incorporated into production schemes will determine the speed at which the goal is achieved of producing livestock more efficiently that meet consumer and market demands.

Microbial Genomics Targeting Food Safety

Advances in biotechnology, microbial genetics, and genomics—especially during the past 5 to 10 years—have greatly affected the development of approaches for assuring the microbial safety of foods, by means of an integrated, farm-to-table framework. Food safety applications of biotechnology and genomics tools have been particularly important in two areas: diagnostic applications, including molecular subtyping methods, and development of animal vaccines for foodborne pathogens. In 1985, invention and subsequent use of the polymerase chain reaction revolutionized biotechnology and many aspects of biology. This method has been applied to develop various foodborne-pathogen detection methods that often are more specific and sometimes more sensitive than traditional culture- or antibody-based detection strategies.Industry and commercial testing laboratories are using PCR kits and detection systems for specific foodborne pathogens. Many PCR assays al low rapid, specific testing for foodborne pathogens and thus enable cost-efficient screening for the presence of foodborne bacteria, viruses, and parasites. In addition to playing an important role in the development of DNA-based detection methods, microbial genomics and biotechnology have had significant impact on improving food safety by providing rapid, discriminatory, and reliable molecular subtyping methods for foodborne pathogens.

Application of these methods has facilitated rapid detection of foodborne disease outbreaks and identification of implicated foods; for example, in 1998–1999, a large foodborne disease outbreak occurred due to L. monocytogenes-contaminated hot dogs and deli meats. In this instance, as well as in other foodborne disease outbreaks occurring in the last few years, molecular subtyping allowed for more rapid outbreak detection and source elimination than would have been possible otherwise.To take further advantage of such molecular subtyping technologies, the Centre for Disease Control and Prevention and state health departments have established PulseNet, a national molecular subtyping network.Implementation of PulseNet in combination with widespread adoption of these new technologies likely will have a positive public health impact by helping in the early detection and control of foodborne disease outbreaks. Application of microbial genomics to vaccine development is covered in the next part of this report, but genetically engineered vaccines for foodborne pathogens have been developed and are available commercially. For example, a genetically engineered Salmonella typhimurium double mutant has been shown to prevent colonization of chickens by both S. typhimurium and S. enteritidis. Further advances in the development of GM vaccines will provide an important opportunity to control foodborne pathogens at the farm level.

USES OF BIOTECHNOLOGY FOR ANIMAL HEALTH AND PRODUCTION

Microbial Genomics Targeting Vaccine Development

Infectious animal diseases continue to rank foremost among the most significant factors limiting efficient production in animal agriculture. Furthermore, infectious agents that are transmitted from animals to humans by way of food and water are an increasing threat to the safety and security of the U.S. food supply and continue to affect human health significantly.

Awareness is increasing that animal agriculture likely will lose the use of several important antimicrobial agents and drug classes for two reasons:

1. Increased resistance among pathogens and
2. Public health threats posed by the potential spread of antimicrobial-resistant zoonotic microbes.

Consequently, new approaches are needed to develop improved tools and strategies for prevention and control of infectious diseases in animal agriculture.

Advances in gene discovery in animal pathogens can be expected to identify new proteins and metabolic pathways, thereby providing a foundation for improved understanding of pathogen biology and ultimately aiding in the design of new and effective therapies. Microbial genomics therefore has the potential not only to identify and to characterize new genes, but also to study the mechanisms governing pathogen development and interactions with their host. Importantly, characterization of genes by means of genomics approaches is a rapid, simple, and cost-effective method of discovering large numbers of new genes and previously unknown biochemical pathways. New disease treatments, whether vaccines or new drugs, must rely on more than empirical methods of discovery and must be based on a fundamental knowledge of pathogen biology and genetics.

The tremendous potential of microbial genomics has led to the development of a focused programme of research on pathogens affecting animals and food safety. Notably, of the more than 70 microbes whose genomes have been elucidated completely thus far, two—Pasteurella multocida and E. coli O157:H7—are of primary importance to animal agriculture. Rapid progress is occurring in sequencing the genomes of many other important food-animal pathogens.An international effort to develop priorities and strategies for animal microbial genomics has been in place since 2000 and has resulted in the development of a list of recommendations for funding agencies and policymakers. In brief, the U. S. Department of Agriculture report has outlined a strategy for identifying the complete sequence of major pathogens of importance to animal health and food safety. The report also highlights the importance of ensuring public access to sequence data to encourage a broad-based effort by the scientific community to use the information for developing the next generation of tools to detect, treat, and prevent infectious diseases and to facilitate understanding of the infectious disease process.

Although much progress has been made in understanding how microbes are pathogenic, advances in microbial genomics of animal pathogens have not as yet resulted in the development of new diagnostic tests, antimicrobial agents, or vaccines of importance for animal agriculture.In the future this likely will change, particularly in regard to the development of diagnostic tests and vaccines. For instance, genomic sequencing of Mycobacterium avium subsp. paratuberculosis, the causative agent of Johne's disease in cattle, has led to the identification of several unique genes that will aid in the specific and sensitive diagnosis of infected cattle, thereby enabling development of rational strategies for controlling the spread of and eventually eradicating the disease. Similarly, a large protein that plays a role in the virulence of Pasteurella multocida has been identified. This protein is a potential target for vaccine development.

ANIMAL GENOMICS FOR PRODUCTION PERFORMANCE AND DISEASE RESISTANCE

Genomic research provides clues to associations between the incidence and severity of infectious bacterial and viral diseases and particular DNA or protein polymorphisms. These associations provide the impetus for whole-genome searches for such loci in production herds. Before 1993, however, the list of definite susceptibility loci remains relatively short. Because no one gene, or no one form of a gene, can confer resistance to all pathogens, even marker-based selection for susceptibility or "resistance" may have to be balanced to ensure that a herd or a population will have an adequate immune response to a variety of pathogens. Alternatively, susceptibility to a specific pathogen, whose presence is confirmed by DNA-based means, could be selected against while maintaining marker and allelic heterozygosity for general immune responsiveness and other production traits.

To identify the genes responsible for genetic variation in host susceptibility to pathogens, genetic markers that account for a significant portion of genetic variance need to be identified and incorporated into a routine screening procedure. A growing number of quantitative trait loci associated with production have been identified as a result of genetic mapping efforts in livestock.

As the human genome effort increases scientific knowledge of the function of genes involved in the complex relationship between host and pathogen, researchers will be able to identify new DNA sequences associated with resistance or susceptibility and to produce genetic markers useful for selection in most, if not all, livestock species. The use of genetic markers as a DNA-based diagnostic tool for parentage testing also can be applied for other diagnostic purposes—for instance, to DNA fingerprinting of a particular inbred line, a cross, or a herd to provide traceability through the feedlot and packing plant to the consumer. Producer and regulator concerns regarding food safety clearly warrant the development of means by which to identify high-quality, biosecure products at the production and retail levels.

The potential for livestock producers to use DNA fingerprints to determine herd identity or to track a product from producer to consumer is real and provides the producer with the added value of genetically brand naming or "bar coding" a product as to its uniqueness or origin. A similar approach can be applied to host susceptibility or resistance to specific pathogens. Scrapie, a member of the transmissible spongiform encephalopathy family that includes bovine spongiform encephalopathy, evidently is transmitted from ewe to lamb at birth or to other ewes exposed to contaminated placental tissues or fluids at lambing.

The disease, which leads to deterioration of the animal's brain, is fatal. Infected sheep show symptoms ranging from frequent rubbing against objects to behavioural changes and uncoordinated movements. Scrapie infects sheep that are positive for the susceptible QQ genotype identified by DNA-based means.

Sheep carrying at least one R gene are believed to be resistant; those carrying both genes seem to be completely resistant. Use of a PCR-based test on DNA from circulating white blood cells to genotype asymptomatic animals for susceptibility status decreases the need for necropsy to identify the presence of the causative agent in brain tissues in the absence of clinical symptoms.

In addition to providing an opportunity for parentage testing, herd or flock identification, and PCRbased diagnostics of infectious pathogens for herd health purposes, DNA-based diagnostics can be used to identify the presence of genetic mutations that diminish livestock production and animal performance and may even be lethal. Once the mutations altering a protein's function are identified, they can be incorporated into DNA-based tests and used to screen parents to determine whether they are carriers of metabolic or other inherited genetic disorders. This screening is accomplished through PCR amplification of the DNA carrying either the normal or the mutated allele and through determination of whether the DNA sequence can be cut at the site of the mutation by an enzyme specific for the DNA sequence in the normal allele.

This technique has been used effectively to identify carriers of inborn errors of metabolism and other phenotypic anomalies in humans as well as in livestock and companion animals.

CONSERVATION OF THE ENVIRONMENT

Impact of Genetically Modified Crops on Animal Manure Management

Meeting the environmental challenges in agriculture is one of the major issues facing the animal industry. Beef and dairy cattle, sheep, swine, and poultry in the United States produce more than 160 million metric tons of manure annually. Most swine and poultry waste is produced in confinement units for which the nearby land base often is insufficient to accommodate waste in an environmentally sound manner.Animal manure, especially that of swine and poultry, is high in nitrogen and phosphorus, both of which can contribute to surface and groundwater pollution. In addition, ammonia and other nitrogenous and sulfurous gases contribute to poor air quality and offensive odors. Several GM crops have been developed or are being developed to address the environmental issues related to N, P, and total manure excretion and odors.

Genetically Modified Crops That Decrease Phosphorus Excretion

Phosphorus content in swine and poultry manure is high because these species consume diets consisting of cereal grains and oilseed meals in which most P is bound organically as phytic acid or phytate. Because of the lack of phytase in their digestive tract, Non-ruminants are unable to degrade phytate, and most P from these feed ingredients is excreted in the feces.

In addition, relatively large amounts of inorganic P must be fed to pigs and poultry to meet their P requirements; consequently, fecal P excretion is increased further. Ruminants use phytate quite efficiently because of the abundance of phytase produced by rumen microorganisms.

The mutant lpal gene in corn interferes with synthesis of phytic acid in corn seeds without affecting total P. Lowphytate corn contains half as much phytate P and greater than three times as much inorganic P as near-isogenic, conventional corn. Recent studies have shown that the bioavailability of P is considerably greater in low-phytate corn than in conventional corn for pigs and for chicks, so less inorganic P is needed in the diet to meet the P requirement when low-phytate corn is used. As a result, 30 to 40% less P is excreted when low-phytate corn is fed to pigs and chicks.

Another biotechnology crop, low-phytate, lowoligosa-ccharide soybean has the potential to decrease further the excretion of P by nonruminants. Soybean meal from low-phytate soybean has less phytate P and more inorganic P than soybean meal from near-isogenic conventional soybean does, and the P in low-phytate soybean meal is more than twice as bioavailable as that in conventional soybean meal for pigs and for chickens. Thus, less P is excreted. Diets containing both lowphytate corn and low-phytate soybean meal have resulted in 50 to 60% decreases in P excretion by pigs and chicks. To date, biotechnology has been used to insert the phytase gene into pigs as well as into alfalfa and canola, and the phytase activity in pigs' saliva and in both crops has been increased greatly as a result.

Genetically Modified Crops That Decrease Nitrogen Excretion

Most N in animal manure originates from dietary sources. Most fecal N comes primarily from dietary protein that is not digested in the digestive tract, and most urinary N arises from absorbed amino acids in excess of animal requirements. Small amounts of excreted N are from the intestinal tract and other endogenous sources.Biotech crops with improved amino acid profiles have the potential to decrease N excretion significantly, especially in Non-ruminants. Increased levels of lysine, methionine, tryptophan, threonine, and other essential amino acids in grains would mean that the essential amino acid requirements of pigs and poultry can be met with lower-protein diets containing fewer excesses of other amino acids that eventually must be degraded to urea N and excreted in the urine.

Moreover, feeding of low-protein diets results in less production of ammonia and other volatile gases from manure that diminish air quality and cause objectionable odors.

Corn containing the mutant opaque-2 or floury-2 gene is an example of a cereal grain having a comparatively high level of lysine, tryptophan, and methionine.

Other cereals have been, or are being, developed with improved amino acid profiles or increased free lysine levels. In the future, transgenic plants will be developed with improved capacities to assimilate N into higherquality, more-digestible proteins with improved amino acid profiles. These products will improve the overall quality of protein consumed by both ruminants and Non-ruminants and will decrease N excretion into the environment.

Genetically Modified Crops That Decrease Total Manure Excretion

Development of new biotechnology crops will affect total manure production. For example, highoil corn was released in the mid-1990s for livestock feeding. Feed intake is decreased when high-energy diets containing high-oil corn are fed and, subsequently, manure output is decreased.

Biotech forages with decreased lignin or with more easily degraded cellulose and hemicellulose are on the horizon. Already, the brown midrib mutant gene, which decreases lignin in corn by as much as 40%, has been bred into certain lines of corn.

Furthermore, lignase, cellulase, and hemicellulase enzymes may be bioengineered into plants to enhance use of these energy sources, thereby decreasing manure output.

Low-oligosaccharide soybeans also have the potential to decrease manure volume as a result of improved energy digestibility in Non-ruminants. In summary, biotechnology offers considerable promise to the animal industry for decreasing nutrients and odors from manure, as well as for decreasing overall manure production. These changes will contribute to a more sustainable environment.

ETHICS IN ANIMAL BIOTECHNOLOGY

Ethical concerns about animal biotechnology usually fall into one or more of four major categories:

1. Mmpacts on human health and safety,
2. Animal welfare and animal rights,
3. Environmental impacts, and
4. Concerns that certain technologies are "unnatural" or amount to "playing God."

Practically and politically, all four categories are important because various interest groups and the public at large are generally concerned about all four, but supporters of animal biotechnologies should address the first two and at least be conversant with issues in the third category. The reason is that the first and second categories are the least problematic, philosophically, insofar as it is relatively easy to articulate a coherent and convincing ethical viewpoint that takes these issues seriously, in sharp contrast to the fourth category. And the third category is either problematic in a similar way or reduces to concern for the first two.

It is difficult to define "unnatural" in a way that both distinguishes the new biotechnologies from traditional agricultural technologies and also makes doing what is unnatural a "bad" thing. Both expressions have strong rhetorical appeal but are probably best interpreted as shorthand ways of expressing concerns that fall into one or more of the other categories.

For instance, when a lobbying group describes a technology as unnatural, the group almost certainly has concerns about risks to humans, animals, and/or the environment, and it is on these concerns that proponents of biotechnology should focus. Concern for the environment may be either "instrumental" or "intrinsic." In the first instance, the concern is that what happens to the environment will in turn affect humans or animals deleteriously. In the second instance, the concern is for the environment such that the environment is itself thought, in some holistic sense, to be an object of direct moral concern.

But in this sense, the moral standing of the environment can be problematic.

Relative to ethics and agricultural biotechnology, the two most important topics for philosophical research are related to:

1. The various kinds of risks posed to humans and
2. The question of how to factor animal well-being into policy decisions.

Regarding the first topic, controversial applications of biotechnology to plant products already have illustrated that public concerns about food safety are deep seated.

Scientists tend to see the application of biotechnology as a purely scientific question, but Comstock (2001) and Thompson (1987) suggest that there are other important conceptual issues in the background. Regarding the second topic, publicity surrounding cloning may heighten public concern about the ways in which scientists factor animal welfare into research decisions. Although cloning may prove to be of little value in agricultural contexts, related media reports constantly quote researchers as stressing that the cloning of humans would be unacceptable in light of the low success rate and the high rate of health problems in animal clones.

This emphasis, by researchers themselves, that cloned animals are likely to have health problems draws increased attention to the questions of how scientists conceive of animal welfare and how they factor adverse welfare impacts on animals into decisions of whether or not various lines of research are justified.

CEASE

Agriculture is transiting a remarkable scientific era with respect to the myriad of novel processes and products that have been developed using biotechnology. These new biotech processes and products already have provided tangible benefits to agriculture and to society. Moreover, many new products of biotechnology are being developed that will benefit the food sector. Development and adoption of novel biotechnologies will be crucial in meeting the challenge of producing enough food for a growing world population while minimizing and reducing impacts on the environment. The impact these technologies have on society in the future, however, will be largely dependent on the extent to which they are adopted by producers and the agricultural community and accepted by consumers.

Questions about societal impacts and safety often arise as the result of technological change. Inherent in the successful development and adoption of new and emerging biotechnologies for agriculture is the need to increase public understanding of the associated scientific, economic, legislative, ethical, and social issues.

10

Dairy Farming for Women

INTRODUCTION

Why don't more females consider dairy farming as a career? It offers the opportunity tomanage a multimillion-dollar business in a dynamic environment. Sometimes your work is like one of a teacher or farm adviser, an HR manager, a vet, a soil scientist, an agronomist, an accountant and the list goes on.

PRESENT SITUATION

A diagram, which illustrates the business plan that my partner Matt Wade and I developed last year to reach our goal of land ownership within eight years.

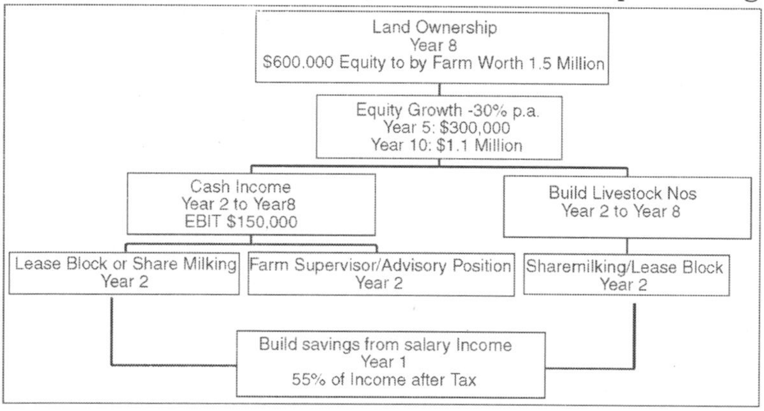

Fig. Strategic Plan Starting 2005 -
Financial and Career Perspective

In June we took on an 800ha lease block in Rotorua as part of our plan. To achieve required equity growth we would build stock numbers and maintain cash income and I would work off farm in a salaried position as a Farm Supervisor/Manager/Advisor. I have recently started as a Farm Consultant working for Farmwise and I am looking forward to meeting lots of Bay of Plenty farmers.

CHALLENGES OF WORKING AS A FEMALE IN THE DAIRY INDUSTRY

- I think that as a female in the industry you need to be one step ahead of the rest – I don't think I would have got the job at Takou Bay Dairy had I not done my degree. In most cases you won't out-muscle them, so out-smart them
- It's easy to get to Herd Manager level. Females are generally great with stock and with people and they are conscientious in their job. Farm owners love these qualities in an employee. It can be more difficult however to get to management level; some farm owners take a little more convincing that a capable female can have the skills necessary to manage the entire operation.
- Learn to laugh - you will still get plenty of situations when it is expected that a male is doing your job. For example, while I as at Takou Bay there were a few times when people would ring up and say, "I'm wanting to talk to the manager – is your husband there? Or worse, "Is your dad there"?
- Some people have trouble believing that a female can have a great farming career and that it's not just a way for her to meet a farmer to marry.
- Having the confidence to manage staff that may not have had a female boss previously.

Knowledge breeds confidence – if you're not confident in a particular area learn as much as you can about it until you are confident.

TIPS FOR OTHER FEMALES CONSIDERING DAIRY FARMING AS A CAREER

- Training is available everywhere. Get as much tuition as possible – that is the easy way to get to management level.
- Interview potential employers, make sure they cut the grade - they need to:
- Be positive and motivated about their farm, the industry and you
- Encourage your career advancement
- Have a performance appraisal system in place
- Provide good working conditions, including good time off, comfortable accommodation, good remuneration and safe working environment
- Be able to teach you what you need to learn
- Be able to explain their farm system and policies to you
- Empower yourself by getting as many practical skills under your belt as possible; know what you need to learn and then find someone capable of teaching you that.
- Look after your body. Get fit before calving starts. When the cows are dry over winter, keep doing exercises to maintain your strength

so you don't hurt your back when calving starts. Have a plan for how you will cope with the extra workload over calving, such as arranging house cleaners, cooks etc. If you are not coping then tell your boss. A good boss will help you to find a solution.

- Don't postpone joy. Make sure you schedule time to do what matters most to you. This can be difficult when you live on the job. While working for Landcorp, I was able to continue with competition on my horses and hunting. I did this by:
- Scheduling personal time on my to do list and onto the work roster
- Training staff to cope in my absence and trusted their ability
- Turning my cell phone off when I was on days off – they could leave a message if it was important
- Having a holiday every year. So as not to conflict with our financial goals, we budget 5% of our earnings each year for travel. As our earnings increase so to will the extent of our travel
- Don't feel that you've got to be good at doing everything. Know your strengths and utilise those and then surround yourself with people whose strengths compliment yours
- Ask people for help – there are so many people in this industry keen to help. Get yourself a mentor!
- Don't let a bad employer put you off dairy farming– for every bad employer there's probably a hundred or more good ones.
- Accept that everyday won't be a good day and be realistic about it - all jobs have their good and bad aspects – dairying is no exception.
- Back yourself and surround yourself with others that back you too.
- And remember, girls can do anything!

WHAT WILL ENCOURAGE MORE FEMALES TO CONSIDER DAIRY FARMING AS A CAREER?

- More television programmes like Macleods Daughters? Not likely. The industry needs to come up with more options for career paths, *i.e.* what happens when farmers want to have a family?
- Not having adverts on fencepost that say, "Wanted young guy for the 06/07 season with a genuine interest in farming".
- Employers maintaining good flexible time off for employees, including during calving. Tools like annual leave and workload planners can be used to ensure that staff numbers are adequate right through the season to provide cover for injuries etc.
- Continuing to profile young women working in the industry through initiatives such as Lets Talk Dairying, 'Window to Dairying', and 'Women in Dairying'.

Bibliography

Amod Sharma and Ravishankar Kumar Singh: *Composite Farming Practices and Economic Development*, Abhijeet Publication, Delhi, 2007.

Anjan Chakrabarti: *Economic Development and Employment in Sikkim*, Authors Press, Delhi, 2009.

Ashutosh Kumar: *Economic Development and Income Distribution*, Deep and Deep Publication, Delhi, 2003.

D.R. Joshi and Sandeep Bajpai: *Economic Development Agricultural and Industrial Growth*, Cyber Tech Publication, Delhi, 2009.

G.S.L.H.V. Prasada Rao, G.G.S.N.Rao and V.U.M. Rao: *Climate Change And Agriculture Over India*, PHI Learning, Delhi, 2003.

Gauri Shanker Dokania and Manisha Dokania: *Economics of Dairy Farming : A Micro Analysis*, Satyam Publishing House, Delhi, 2012.

Golam Kibria, A.K. Yousuf Haroon, Dayanthi Nugegoda and Gavin Rose: *Climate Change and Chemicals : Environmental and Biological Aspects*, New India Publishing Agency, Delhi, 2010.

Harmeet Singh: *Dairy Farming*, APH Publication, Delhi, 2011.

Hema Khanna and G.S. Bhalla: *Economic Development and Ecological Degradation*, Regal Publications, Delhi, 2010.

L C Mallaiah and L Ratna Kumari: *Economic Development and Dalit Agricultural Workers*, Abhijeet Publication, Delhi, 2007.

M. Singh, D.K. Sharma and U.K. Mishra: *Organic Dairy Farming*, Satish Serial Publication, Delhi, 2011.

M. Vasu: *Economic Development and Poverty*, Abhijeet Publication, Delhi, 2009.

Man Chand Khandela: *Economic Development : Myth and Reality*, Pointer Publication, Delhi, 2007.

Manoj Negi: *Career in Dairy Farming*, Abhishek Publication, Delhi, 2010.

Md. Mukhtar Alam: *Economic Development and Contemporary Geopolitics : The Indian Ocean Rim*, Om Publication, Delhi, 2004.

Misra and Puri: *Economic Development and Policy in India*, Himalaya Publishing House, Delhi, 2010.

Mohd. Kalimuddin: *Milk Analysis and Dairy Farming*, Enkay Publication, Delhi, 2012.

N. Kumar and R. Mittal: *Economic Development and Planning*, Anmol Publication, Delhi, 2002.

Nanda Nautiyal, Vir Singh, Neelendra K. Joshi and Akanksha Rastogi: *Climate Change and Biodiversity*, Biotech Books, Delhi, 2012.

P.A. Koli: *Economic Development and Environment Issues*, Serials Publication, Delhi, 2005.

Profulla Hazarika: *Economic Development and Ecological Balance in Assam*, D.V.S. Publication, Delhi, 2010.

Raman Singh: *Economic Development and Planning*, ABD Publication, Delhi, 2011.

S.D. Chamola: *Economic Development and Ethics : Indian Experience Since Independence*, Global Vision Publication, Delhi, 2011.

S.D. Singh: *Economic Development*, Centrum Press, Delhi, 2010.

Shyam Kishore Singh: *Dairy Farming*, Alfa Publications, Delhi, 2012.

Thomas E Lovejoy and Lee Hannah: *Climate Change and Biodiversity*, TERI Publication, Delhi, 2006.

Uma Shankar Singh: *Dairy Farming*, Anmol Publication, Delhi, 2008.

V. Nath: *Economic Development and Planning in India*, Concept Publication, Delhi, 2010.

Vir Singh and Babita Bohra: *Dairy Farming in Mountain Areas*, Daya Publication, delhi, 2006.

Y.V.R. Reddy, M. Jayachandra Reddy, B. Hemalatha and Y.S. Ramakrishna: *Economic Evaluation of Dairy Farming in Rural India*, Agrotech Publication, Delhi, 2006.

Index

L

M

N

P

R

S

T

U

V

W